Spatial analysis and GIS

**Technical Issues in
Geographic Information Systems**

Series Editors:

Donna J. Peuquet, The Pennsylvania State University
Duane F. Marble, The Ohio State University

Also in this series:

Gail Langran, *Time in GIS*

Spatial analysis and GIS

Edited by

Stewart Fotheringham and **Peter Rogerson**
Department of Geography, SUNY at Buffalo

Taylor & Francis
Publishers since 1798

UK Taylor & Francis Ltd, 4 John St, London WC1N 2ET
USA Taylor & Francis Inc., 1900 Frost Road, Suite 101, Bristol PA 19007

British Library Cataloguing in Publication Data

A catalogue record for this book is available from the British Library

ISBN 0 7484 0103 2 (cased)
      0 7484 0104 0 (paper)

Library of Congress Cataloging in Publication Data are available

Cover design by Amanda Barragry

Typeset by Graphicraft Typesetters Ltd., Hong Kong

Printed in Great Britain by Burgess Science Press, Basingstoke on paper which has a specified pH value on final paper manufacture of not less than 7.5 and is therefore 'acid free'.

Contents

Contributors   vii

Acknowledgements   ix

1   GIS and spatial analysis: introduction and overview   1
Peter A. Rogerson and A. Stewart Fotheringham

PART I   INTEGRATING GIS AND SPATIAL ANALYSIS: AN OVERVIEW OF THE ISSUES   11

2   A review of statistical spatial analysis in geographical information systems   13
Trevor C. Bailey

3   Designing spatial data analysis modules for geographical information systems   45
Robert Haining

4   Spatial analysis and GIS   65
Morton E. O'Kelly

PART II   METHODS OF SPATIAL ANALYSIS AND LINKAGES TO GIS   81

5   Two exploratory space-time-attribute pattern analysers relevant to GIS   83
Stan Openshaw

6   Spatial dependence and heterogeneity and proximal databases   105
Arthur Getis

7   Areal interpolation and types of data   121
Robin Flowerdew and Mick Green

8   Spatial point process modelling in a GIS environment   147
Anthony Gatrell and Barry Rowlingson

9   Object oriented spatial analysis   165
Bruce A. Ralston

PART III   GIS AND SPATIAL ANALYSIS: APPLICATIONS   187

10   Urban analysis in a GIS environment: population density modelling using ARC/INFO   189
Michael Batty and Yichun Xie

Contents

11 Optimization modelling in a GIS framework:
    the problem of political redistricting   221
    Bill Macmillan and T. Pierce

12 A surface model approach to the representation of
    population-related social indicators   247
    Ian Bracken

13 The council tax for Great Britain: a GIS-based
    sensitivity analysis of capital valuations   261
    Paul Longley, Gary Higgs and David Martin

    Index   277

Contributors*

T. R. Bailey
Department of Mathematical Statistics and Operational Research, University of Exeter, Laver Building, North Park Road, UK

Michael Batty
NCGIA/Department of Geography, University at Buffalo, Buffalo, NY 14261, USA

Ian Bracken
Department of City and Regional Planning, University of Wales, Cardiff, P.O. Box 906, Colum Drive, Cardiff CF1 3YN, UK

Robin Flowerdew
Department of Geography, University of Lancaster, Lancaster LA1 4YB, UK

Stewart Fotheringham
NCGIA/Department of Geography, University at Buffalo, Buffalo, NY 14261, USA

Tony Gatrell
Department of Geography, University of Lancaster, Lancaster LA1 4YB, UK

Art Getis
Department of Geography, San Diego State University, San Diego, CA 92182, USA

Bob Haining
Department of Geography, University of Sheffield, Sheffield S10 2TN, UK

Paul Longley
Department of Geography, University of Bristol, University Road, Bristol BS8 1SS, UK

Bill Macmillan
School of Geography, Oxford University, Mansfield Road, Oxford OX1 3TB, UK

Morton O'Kelly
Department of Geography, The Ohio State University, 103 Bricker Hall, 190 North Oval Mall, Columbus, OH 43210–1361, USA

Stan Openshaw
School of Geography, University of Leeds, Leeds, LS2 9JT, UK

---

* First named authors only

Bruce Ralston
Department of Geography, University of Tennessee, 408 G and G Building, Knoxville, TN 37996–1420, USA

Peter Rogerson
NCGIA/Department of Geography, University at Buffalo, Buffalo, NY 14261, USA

Acknowledgements

The chapters in this book were originally prepared for a Specialist Meeting of the National Center for Geographic Information and Analysis (NCGIA) on GIS and Spatial Analysis. We wish to thank Andrew Curtis, Rusty Dodson, Sheri Hudak and Uwe Deichmann for taking detailed notes during that meeting. We also thank Andrew Curtis and Connie Holoman for their assistance in preparing a written summary of the meeting. Both the notes and the summary were especially valuable in preparing the introductory chapter of this book. Connie Holoman and Sandi Glendenning did a tremendous job in helping to organize the meeting and we owe them a great debt. We are also grateful to the National Science Foundation for their support to the NCGIA through grant SES-8810917 and to the financial support of the Mathematical Models Commission of the International Geographical Union.

To Neill and Bethany

1

GIS and spatial analysis: introduction and overview

Peter A. Rogerson and A. Stewart Fotheringham

History of the NCGIA initiative on GIS and spatial analysis

A proposal for a National Center for Geographic Information and Analysis (NCGIA) Initiative on Geographic Information Systems (GIS) and Spatial Analysis was first submitted to the Scientific Policy Committee of the NCGIA in March 1989. It was formally resubmitted in June 1991 after being divided into separate proposals for initiatives in 'GIS and Statistical Analysis' and 'GIS and Spatial Modeling'. The essence of the former of these two proposals was accepted and evolved into the more generic 'GIS and Spatial Analysis' initiative that was approved, with the expectation that an initiative emphasizing spatial modelling would take place at a later date.

The contributions in this book were originally prepared for the Specialist Meeting that marked the beginning of the NCGIA Initiative on GIS and Spatial Analysis. The Specialist Meeting was held in San Diego, California, in April 1992, and brought together 35 participants from academic institutions, governmental agencies, and the private sector. A list of participants is provided in Table 1.1. A facet of the initiative conceived at an early stage was its focus on substantive applications in the social sciences. There is perhaps an equally strong potential for interaction between GIS and spatial analysis in the physical sciences, as evidenced by the sessions on GIS and Spatial Analysis in Hydrologic and Climatic Modeling at the Association of American Geographers Annual Meeting held in 1992, and by the NCGIA-sponsored meeting on GIS and Environmental Modeling in Colorado.

The impetus for this NCGIA Research Initiative was the relative lack of research into the integration of spatial analysis and GIS, as well as the potential advantages in developing such an integration. From a GIS perspective, there is an increasing demand for systems that 'do something' other than display and organize data. From the spatial analytical perspective, there are advantages to linking statistical methods and mathematical models to the database and display capabilities of a GIS. Although the GIS may not be absolutely necessary for spatial analysis, it can facilitate such analysis and may even provide insights that would otherwise be missed. It is possible, for example, that the representation of spatial data and model results within a

Table 1.1. Specialist Meeting participant list.

T. R. Bailey
Department of Mathematical Statistics
and Operational Research
University of Exeter

Michael Batty
NCGIA/Department of Geography
SUNY at Buffalo

Graeme Bonham-Carter
Geological Survey of Canada
Energy, Mines, and Resources

Ian Bracken
Department of City and Regional
Planning
University of Wales, Cardiff

Ayse Can
Department of Geography
Syracuse University

Noel Cressie
Department of Statistics
Iowa State University

Andrew Curtis
NCGIA/Department of Geography
SUNY at Buffalo

Lee De Cola
United States Geological Survey
National Mapping Division

Paul Densham
NCGIA/Department of Geography
SUNY at Buffalo

Uwe Diechmann
NCGIA
University of California, Santa Barbara

Rusty Dodson
NCGIA
University of California, Santa Barbara

Randall Downer
Applied Biomathematics

Robin Dubin
Department of Economics
Case Western Reserve University

Chuck Ehlschlaeger
US Army Construction Engineering
Research Lab

Manfred M. Fischer
Department of Economic and Social
Geography
Vienna University of Economics and
Business Administration

Robin Flowerdew
Department of Geography
University of Lancaster

Stewart Fotheringham
NCGIA/Department of Geography
SUNY at Buffalo

Tony Gatrell
Department of Geography
University of Lancaster

Art Getis
Department of Geography
San Diego State University

Michael Goodchild
NCGIA/Department of Geography
University of California, Santa Barbara

Bob Haining
Department of Geography
University of Sheffield

Dale Honeycutt
Environmental Systems Research
Institute

Sheri Hudak
NCGIA
University of California, Santa Barbara

Clifford Kottman
Intergraph Corporation

Paul Longley
Department of City and Regional
Planning
University of Wales at Cardiff

Bill Macmillan
School of Geography
Oxford University

Morton O'Kelly
Department of Geography
The Ohio State University

Stan Openshaw
School of Geography
University of Leeds

Table 1.1. (cont.)

Bruce Ralston
Department of Geography
University of Tennessee

Peter Rogerson
NCGIA/Department of Geography
SUNY at Buffalo

Peter Rosenson
Geographic Files Branch
U.S. Bureau of the Census

Gerard Rushton
Department of Geography
San Diego State University

Howard Slavin
Caliper Corporation

Waldo Tobler
NCGIA/Department of Geography
University of California, Santa Barbara

Roger White
Department of Geography
Memorial University of Newfoundland

GIS could lead to an improved understanding both of the attributes being examined and of the procedures used to examine them. It is in this spirit that we have collected a set of papers presented at the meeting, which we feel lead the way in describing the potential of GIS for facilitating spatial analytical research.

The objectives of the initiative are in keeping with the aims of the National Center, as identified in the original guidelines from the National Science Foundation. The original solicitation for a National Center for Geographic Information and Analysis circulated by the National Science Foundation in 1987 contained as one of its four goals to 'advance the theory, methods, and techniques of geographic analysis based on geographic information systems in the many disciplines involved in GIS research' (National Science Foundation, 1987, p. 2). The solicitation also notes that the research program of the NCGIA should address five general problem areas, including 'improved methods of spatial analysis and advances in spatial statistics'.

GIS and spatial analysis

Geographic information systems were initially developed as tools for the storage, retrieval and display of geographic information. Capabilities for the geographic analysis of spatial data were either poor or lacking in these early systems. Following calls for better integration of GIS and the methods of spatial analysis (see, for example, Abler, 1987; Goodchild, 1987; National Center for Geographic Information and Analysis, 1989), various alternatives have now been suggested for such an integration (Openshaw, 1990; Haining and Wise, 1991; Rowlingson *et al.*, 1991). As Fotheringham and Rogerson (1993) note, 'progress in this area is inevitable and ... future developments will continue to place increasing emphasis upon the analytical capabilities of GIS'.

Consideration of the integration of spatial analysis and GIS leads naturally to two questions: (1) how can spatial analysis assist GIS, and (2) how can GIS assist spatial analysis? Under these general headings, a myriad of more specific questions emerges. The following are representative (but not exhaustive) specific questions given to participants prior to the specialist meeting:

1. What restrictions are placed on spatial analysis by the modifiable areal unit problem and how can a GIS help in better understanding this problem?
2. How can GIS assist in exploratory data analysis and in computer-intensive analytical methods such as bootstrapping and the visualization of results?
3. How can GIS assist in performing and displaying the results of various types of sensitivity analysis?
4. How can the data structures of a GIS be exploited in spatial analytical routines?
5. What are the important needs in terms of a user interface and language for spatial analysis performed on a GIS?
6. What are some of the problems in spatial analysis that should be conveyed to a GIS user and how should these problems be conveyed?

The papers prepared for the Specialist Meeting reflect a concern for questions such as these. More generally, they represent a cross-section of the multifarious issues that arise in the integration of GIS and spatial analysis.

After reading the conference papers it became clear to us that each could be placed into one of three categories; consequently, we have divided up this book into three parts that correspond to these categories. Part I contains three chapters – by Bailey, Haining and O'Kelly – that focus directly upon some of the broader issues associated with the integration of spatial analysis and GIS, and the authors provide useful and valuable perspectives.

In Chapter 2, Bailey evaluates the progress that has been made in using GIS technology and functionality for problems requiring spatial statistical analysis. He also assesses the potential for future developments in this area. Bailey achieves this difficult task by first dividing the methods of spatial statistical analysis into eight classes. The eight classes are chosen to reflect the dimensions of the methods that are relevant to GIS, and indeed it is the success of this classification scheme that allows Bailey successfully to assess progress and potential. Bailey concludes that there is substantial potential for furthering the contributions of GIS to spatial statistical analysis, particularly in the areas of simple, descriptive statistics and the analysis of covariance.

In Chapter 3, Haining provides a complementary assessment of the interface between GIS and spatial statistical analysis. He reiterates the value of linking spatial statistical analysis and GIS, and argues for GIS developments that aid both exploratory and confirmatory modes of analysis. By using an application to the study of the intra-urban variation in cancer mortality rates, Haining suggests that the following six questions must be addressed prior to the successful linkage between spatial statistical analysis and GIS:

1. What types of data can be held in a GIS?
2. What classes of questions can be asked of such data?
3. What forms of statistical spatial data analysis (SDA) are available for tackling these questions?
4. What minimum set of SDA tools is need to support a coherent program of SDA?
5. What are the fundamental operations needed to support these tools?
6. Can the existing functionality within GIS support these fundamental operations?

Haining concludes that GIS provides a valuable means for making statistical spatial data analysis accessible to the user, and that it is now important to focus upon the scope of SDA tools that should go into GIS software, as well as the GIS functionality that will be required to support the tools.

Several important themes also emerge from Chapter 4. O'Kelly begins by suggesting that attention be given to two major directions: (1) the improvement of traditional methods for displaying, exploring, and presenting spatial data, and (2) the need to help GIS users understand and improve the methods of spatial analysis that they are using. O'Kelly argues that benefits to integration can accrue in both directions. Thus GIS can be enhanced through the addition of spatial analysis functions, and spatial analysis functions may potentially be improved through their use in GIS. O'Kelly emphasizes these points throughout his paper with applications to space-time pattern recognition, spatial interaction, spatial autocorrelation, and the measurement of spatial situation. O'Kelly raises a number of issues that will clearly be relevant and important as the methods of spatial analysis are made a part of GIS functionality. These include:

1. The simultaneous difficulty and importance of finding spatial, temporal, or space-time patterns in large spatial databases.
2. The production of large databases raises a number of issues. How can the barriers to users generated by the complexity and size of many databases be reduced? How can the quality of the data be assessed? How can the data be used, even in light of inaccuracies? How might novel approaches to visualization be useful in addressing some of these questions?
3. Might the recent improvements in computational and GIS technology generate a renaissance for particular methods of spatial analysis, such as point pattern analysis?

O'Kelly concludes by emphasizing that geographers need to play more of a role in ensuring 'the timely and accurate usage of sophisticated spatial analysis tools' in a GIS environment.

Part II of the book contains chapters that are primarily oriented towards either specific methods of spatial analysis or the linkages between spatial analysis and GIS. In Chapter 5 Openshaw argues that exploratory methods of spatial analysis are ideally suited for GIS, and should be given greater

emphasis. He focuses upon the need for developing pattern detectors that
(1) are not scale specific, (2) are highly automated, and (3) have the flexibility
of including human knowledge. He suggests that previous attempts to find
spatial pattern, temporal pattern, space-time interaction, and clusters of highly
correlated attributes are too limited, and that all three should be viewed
simultaneously in 'tri-space'. Openshaw offers two types of pattern analyzers
– the first is an algorithm based upon grid search and Monte Carlo significance tests, and the second involves the genetic evolution of pattern detectors through the survival of good detectors, and the death or weeding out of
poor ones. Openshaw demonstrates the approaches through an application to
crime pattern detection. He concludes by reiterating the need for methods
that explore GIS databases in a way that allows such activity to be 'fluid,
artistic, and creative', and in a way that stimulates 'the user's imagination and
intuitive powers'.

In Chapter 6 Getis also extols the virtues of exploratory data analysis, and
describes how the traditional geographic notions of 'site' and 'situation' may
be integrated and used to evaluate the degree of heterogeneity in spatial data
bases. He advocates the inclusion of measures of spatial dependence in data
sets. This would result in many benefits, including the evaluation of scale
effects and in the identification of appropriate models. Getis makes these
suggestions in the broader context that within GIS, space must ultimately be
viewed as more than a container of locations devoid of situational context.
The view that to be useful, GIS must incorporate relational views of space
as well as absolute ones, has also been forcefully made by Couclelis (1991).
Getis takes an important step towards addressing the issue.

In Chapter 7 Flowerdew and Green attack a very practical problem that
faces spatial analysts all too often. Spatial data are often needed for spatial
units other than those for which they are collected. Some type of interpolation
is inevitably required to produce the data for the required geographic units
(dubbed 'target zones' by Flowerdew and Green) from the units for which
data are available ('source zones'). Flowerdew and Green first describe an
approach to areal interpolation that makes use of ancillary data. The value
of a variable for a particular target region is estimated via the relationship
between that variable and an ancillary variable, as well as the known value
of the ancillary variable for the target region. Flowerdew and Green also
provide details regarding the implementation of the approach within ARC/
INFO, as well as the obstacles encountered along the way.

In Chapter 8 Gatrell and Rowlingson also focus discussion upon the linkages between GIS and spatial analysis. In particular, they address the issues
that arise when ARC/INFO is used for point pattern description and for
estimating a spatial point process model. Gatrell takes two separate approaches
– the first is to link FORTRAN routines for point pattern analysis with ARC/
INFO, and the second is to develop the appropriate routines within S-Plus,
using a set of functions known as SPLANCS that have been developed
specifically for point pattern analysis. Gatrell concludes that much of what is

accomplished by the effort is 'to do within a GIS environment what spatial statisticians have been doing for 15 years'. This indeed seems to be the status of many of the other efforts along these lines (for example, see the chapters by Flowerdew and Green, and Batty and Xie). It has now been demonstrated, through a good deal of effort, that various forms of spatial analysis can indeed be carried out in a GIS environment. The challenge now is to take full advantage of the capabilities of GIS in furthering the development and use of the methods of spatial analysis.

In the last chapter of Part II, Ralston presents the case for an object-oriented approach to spatial analysis in GIS. He argues that an object-oriented programming approach facilitates the development of the appropriate re-usable tools that form the building blocks of spatial modelling. In addition, the approach is a more natural one when the problem facing the spatial analyst is either modified or made more complex, since it forces the analyst to think more about the elements of the problem at hand and less about the programming changes that are necessitated. Ralston illustrates these points through an application to optimization problems.

Part III of the book contains four chapters that focus directly upon issues associated with spatial modelling and GIS. In Chapter 10 Batty and Xie describe how population density models may be embedded within ARC/INFO. Their focus is not upon the population density models themselves, but rather upon the development of a prototypical system that is effective in demonstrating how such modelling may be carried out in a GIS environment. Batty and Xie use ARC/INFO to develop both a user interface that facilitates the interaction between the system and the user, and an interface that facilitates the transitions between data description, data display, model estimation and projection. They view their contribution as an exercise designed to push the limits of conventional GIS as far as possible to allow inclusion of traditional planning models. An alternative strategy to the adopted one of staying within the confines of an 'off-the-shelf' GIS would be to start with the model itself, and build in the needed GIS functionality. Batty and Xie do not explore this latter strategy. They do, however, make the interesting point that when spatial analysis and GIS are strongly coupled together (i.e. with integrated software for performing model and GIS functions, and little or no passing of input files to the software during program execution), there is the possibility of having some systems that are more oriented towards modelling and some that are more oriented towards GIS capabilities. Batty and Xie cite Ding and Fotheringham's (1992) Spatial Analysis Module as an example of a system more oriented towards analysis and interpretation, while their own model of the Buffalo region is more oriented toward GIS, since visualization and display play such a central role.

In Chapter 11, Macmillan and Pierce begin by noting that the quantitative revolution in geography produced many methods which, though seemingly holding much promise at the time, ended up contributing little. They suggest that by using GIS to rebuild some of these models, perhaps additional returns may be achieved. The main task Macmillan and Pierce set for themselves is to describe in detail how a simulated annealing approach to solving political redistricting problems can be used within GIS (in particular, within TransCAD). In this regard, the chapter is similar to that of Batty and Xie, since a specific modelling problem is tied to a specific GIS. However, Macmillan and Pierce are more concerned with the modelling than with the display and the interface. As they point out, systems such as this are more sophisticated than more common 'passive' systems that, in the case of redistricting, allow the user to interact with a given plan through a trial-and-error like process of adding, modifying or deleting the subregions associated with particular districts. It is important that spatial analysts make similar contributions to the development of GIS applications; otherwise, users will continue simply to use the passive systems without having the opportunity to use more sophisticated methods that rely more heavily on spatial analysis.

In Chapter 12, Bracken addresses the problems associated with the analysis of data collected for geographic regions by developing a surface model to represent the data. The intent of the model is to represent population and other related variables 'using a structure which is independent of the peculiar and unique characteristics of the actual spatial enumeration'. Bracken's specific aim is to transform both zone-based and point-based data into a representation that approaches a continuous surface. Ultimately, he ends up with data that are mapped onto a fine and variable resolution grid. He achieves this by estimating the population of cell i as a weighted average of the recorded populations at points j, where the weights are determined by the strength of the connection between cell i and point j. Weights are determined by a distance decay function, with weights associated with pairs separated by more than a given distance set equal to zero. Bracken's principal contribution is to 'provide a form of data representation that is inherently more suitable to the display, manipulation, and portrayal of socioeconomic information', and to facilitate the integration and analysis of multiple sources and types of data by using a common geographical structure.

In the final chapter, Longley, Higgs, and Martin use GIS to consider the spatial impacts of a local council tax on households in Cardiff. The asking prices of houses for sale were first determined, and were then used to estimate the capital value for all dwellings. The authors demonstrate the utility of GIS in examining alternative scenarios, and they illustrate this flexibility by assessing the effects of a 10 per cent decrease in valuation.

Summary

The chapters contained in this volume represent both a statement of where research presently stands in terms of the relationship between GIS and spatial analysis and where research ought to be headed. Several general themes appear to emerge from these discussions. One concerns the relationship between GIS and exploratory and confirmatory modes of analysis, which is the subject of commentary elsewhere (Fotheringham, 1992). The adjective 'exploratory' usually describes those analytical methods where the results *suggest* hypotheses, while 'confirmatory' analyses are used to *test* hypotheses, although the distinction between the two is at times fuzzy and there are several types of statistical analysis that could fall into both areas. There was a general consensus at the meeting that although connections between GIS and confirmatory statistical packages have been established, there was greater potential for new insights in the combination of GIS and exploratory techniques. GIS are data rich and contain excellent display capabilities; exploratory data analysis is data-hungry and generally visual. It was generally felt at the meeting that real gains in exploratory spatial data analysis could result from the integration with GIS and Fotheringham (1993) describes several specific areas of research that could profit from this integration.

A second general theme is that of the relationship between GIS and the development of geographic theory. Much empirical based research suggests theory or tests it, both actions being integral components of the development process. GIS should be seen as a tool that can assist in the development of geographic theory through facilitating empirical research. The integration of GIS and spatial analysis is aimed therefore at only a subset of spatial analysis: that which deals with applied spatial modelling and with empirical analysis. Within those limits, the technology should prove extremely useful. There is an opportunity to utilize the power of GIS technology to help understand some basic geographic problems such as the sensitivity of analytical results to zone definition, the nature of spatial nonstationarity, and the definition of spatial outliers. Most of the authors in this book would agree that we should be careful not to be carried away, however, with the power of the technology GIS affords so that theoretical research takes second place.

Finally, it remains to be seen what insights into the analysis of spatial data will be generated by the access to excellent display capabilities, database operations and spatial querying facilities a GIS provides. The chapters in this book signal the way in which these insights might be gained and what some of them might be. The next decade should see a surge in interest in spatial analysis within geography and other academic disciplines, as well as in the private sector. It is therefore inevitable that geographic information systems will have increasingly sophisticated spatial analytical capabilities; this book serves to signal what lies ahead. It is perhaps fair to say that we have to this point spent a large amount of time 'reinventing the wheel', that is, getting methods that are already operational running in a GIS environment. It is now time, and the future seems promising, to go beyond this to use the capabilities of GIS to enhance models and methods and to make them more efficient.

References

Abler, R., 1987, The National Science Foundation National Center for Geographic Information and Analysis, *International Journal of Geographical Information Systems*, **1**, 303–26.

Couclelis, H., 1991, Requirements for planning-related GIS: a spatial perspective, *Papers in Regional Science*, **70**, 9–19.

Fotheringham, A. S., 1992, Exploratory spatial data analysis and GIS Commentary, *Environment and Planning A*, **24**(12), 1675–78.

Fotheringham, A. S. and Rogerson, P. A., 1993, GIS and spatial analytical problems, *International Journal of Geographical Information Systems*, **7**(1), 3–19.

Goodchild, M., 1987, A spatial analytical perspective on geographical information systems, *International Journal of Geographical Information Systems*, **1**, 327–34.

Haining, R. P. and Wise, S. M., 1991, GIS and spatial analysis: report on the Sheffield Workshop, Regional Research Laboratory Initiative Discussion Paper 11, Department of Town and Regional Planning, University of Sheffield.

National Center for Geographic Information and Analysis, 1989, The research plan of the National Center for Geographic Information and Analysis, *International Journal of Geographic Information Systems*, **3**, 117–36.

National Science Foundation, 1987, National Center for Geographic Information and Analysis, Directorate for Biological, Behavioral, and Social Sciences, Guidelines for Submitting Proposals.

Openshaw, S., 1990, A spatial analysis research strategy for the regional research laboratory initiative, Regional Research Laboratory Initiative Discussion Paper 3, Department of Town and Regional Planning, University of Sheffield.

Rowlingson, B. S., Flowerdew, R. and Gatrell, A., 1991, Statistical spatial analysis in a geographical information systems framework. Research Report 23, North West Regional Laboratory, Lancaster University.

PART I

Integrating GIS and spatial analysis:
an overview of the issues

2

A review of statistical spatial analysis in geographical information systems

Trevor C. Bailey

Introduction

Despite widespread recognition that the analysis of patterns and relationships in geographical data should be a central function of geographical information systems (GIS), the sophistication of certain areas of analytical functionality in many existing GIS continues to leave much to be desired. It is not the objective of this chapter to labour this point. The problem has been widely acknowledged – the importance of the identification of relevant spatial analysis tools and their links to GIS was mentioned in the eponymous Chorley report, *Handling Geographical Information* (Department of the Environment, 1987), and subsequently appears as a key issue in the research agendas of both, the US National Centre for Geographical Information and Analysis (NCGIA, 1989), and the UK joint ESRC/NERC initiative on GIS (Masser, 1988). The same theme has constantly recurred in the GIS literature (e.g. Goodchild, 1987; Rhind, 1988; Rhind and Green, 1988; Openshaw, 1990; Burrough, 1990; Haining and Wise, 1991; Anselin and Getis, 1991).

The intention of this chapter is to review and comment on the progress in this area of GIS research. More specifically, the chapter concentrates on a particular aspect of that research – the linkage between GIS and methods for the *statistical analysis* of spatial data. This is felt to be a subset of analytical functionality within GIS which offers considerable potential for development and which is of sufficiently general interest to the GIS community to merit special attention.

A review of this area is felt timely for two reasons. Firstly, there is *increased interest* – in one sense, GIS technology is now beginning to reach the stage where a number of users are beginning to mature. Initial problems in establishing a spatial database and gaining a familiarity with their chosen GIS have largely been overcome, and users are now beginning to grapple with the analysis of patterns in spatial data and with investigating possible explanations for these. The result is a generally increased interest in which spatial analysis methods might be appropriate for various types of investigation and in whether, or how, they may be used in a GIS environment. Secondly, there is *increased activity* – one only has to look at the number of practical case studies reported in the GIS literature which involve spatial analysis to appreciate that people are finding ways to perform a variety of spatial analyses in conjunction with

GIS, albeit with some difficulties. Some researchers have been able to use the functions which exist within existing commercial GIS. ARC-INFO, for example, includes functions for location/allocation modelling and gridding which facilitate some forms of statistical analysis (ESRI, 1990) and also includes options for kriging. Others have demonstrated what can be achieved by linking GIS to existing statistical packages (e.g. Kehris, 1990a; Ding and Fotheringham, 1991); or have attempted to develop software which allows a dynamic link between mapping and analysis, and opens up a whole new range of opportunities in the understanding of spatial relationships (Haslett et al., 1990). Yet others have suggested and implemented entirely new methods of analysis for use in a GIS environment (Openshaw et al., 1987; Openshaw et al., 1990); or have addressed themselves to the question of what should constitute the basic functional requirements of a 'language' for spatial analysis and to defining a corresponding research agenda to implement such ideas (e.g. Goodchild, 1987; Dixon, 1987; Openshaw, 1991).

All this has resulted in a fairly substantial body of literature which concerns the interface between GIS and spatial analysis, together with a somewhat confusing range of software packages and links, each supporting different analysis functions. It is therefore felt both timely and helpful to attempt to present a coherent picture of the field and try to crystallize some of the key issues involved. This chapter will certainly not be the only such review – recently various research groups interested in spatial analysis in GIS have begun to come together to share experiences and discuss how the techniques of spatial analysis should be linked to GIS. This has led to a number of publications which review aspects of the field. A workshop on the subject has been held in Sheffield, United Kingdom, last year (Haining and Wise, 1991) and a lengthier discussion of some of the topics raised there published (Goodchild et al., 1992). Openshaw has discussed similar material independently (Openshaw, 1990; Openshaw, 1991; Openshaw et al., 1991), as have, Anselin and Getis (1991), Ding and Fotheringham (1991), Wise and Haining (1991), and Rowlingson et al. (1991). It is hoped that the viewpoints expressed in this chapter will help to clarify further some of the issues involved and will add to and inform this general debate.

The approach adopted in the chapter is, firstly, to clarify the distinction between spatial analysis in general and statistical spatial analysis in particular. Then, from the range of existing techniques of statistical spatial analysis, the intent is to identify those which are thought to be the most potentially useful to discuss in relation to GIS and to broadly classify these into their major application areas. In order to counteract a tendency to produce a 'shopping list' of techniques, those selected are then grouped for subsequent discussion. Secondly, to identify the general benefits involved in close interactive linkage between spatial statistics and the sort of functions which GIS provide for geographical visualisation, spatial query and the derivation of spatial relationships. Thirdly, I analyse more specifically what these mean in respect of each of the groups of statistical methods previously identified and, at the same time, to discuss what progress has been made in realising these benefits – either within existing GIS, or within GIS-related products, or in the various combinations of GIS and other modelling or statistical packages suggested by various researchers. Finally, to attempt to summarize the relationship between potential and progress in each of the defined areas, to discuss general issues in the field and to suggest what future developments may be valuable.

Potentially useful statistical spatial analysis techniques in relation to GIS

One difficulty experienced in any discussion of links between GIS and spatial analysis is clarification of exactly what is to be considered as spatial analysis. The problem arises because, by its nature, GIS is a multi-disciplinary field and each discipline has developed a terminology and methodology for spatial analysis which reflects the particular interests of that field. In the face of such a diversity of analytical perspectives, it is difficult to define spatial analysis any more specifically than as: *'a general ability to manipulate spatial data into different forms and extract additional meaning as a result'*.

However, this chapter restricts the discussion to statistical spatial analysis – methods which address the inherent stochastic nature of patterns and relationships, rather than forms of analysis which are purely deterministic. The emphasis on statistical spatial analysis should not be taken to imply that the provision of other areas of analytical functionality is not equally important, such as those that arise from network analysis, routing, transportation, location/allocation modelling, site selection, three-dimensional modelling and projection or cartographic algebra. But simply, that such forms of analysis are in general better catered for, although not without deficiencies, in existing GIS than their statistical counterparts. For example ARC/INFO (ESRI, 1990) offers network analysis functions for calculation of spanning trees and shortest paths, and an ALLOCATE function which enables certain forms of deterministic location/allocation modelling. Links to more sophisticated algorithms for solution to 'travelling salesman' problems have proved possible (Vincent and Daly, 1990), three-dimensional visualization, projection, and calculation of slopes and aspect, are also standard in most GIS.

A further distinction is made in this chapter between the spatial *summarization* of data and the spatial *analysis* of such data. The former is taken to refer to basic functions for the selective retrieval of spatial information within defined areas of interest and the computation, tabulation or mapping of various basic summary statistics of that information. The second is more concerned with the investigation of patterns in spatial data – in particular, in seeking possible relationships between such patterns and other attributes or features within the study region, and with the modelling of such relationships for the

purpose of understanding or prediction. It is widely acknowledged that existing GIS systems offer a powerful array of techniques for spatial *summarization*
(query facilities allow flexible data retrieval, Boolean operations are provided
on the attributes of points, lines and polygons, as are techniques for line
intersection, point-in-polygon or polygon overlay, addition and subtraction of
map layers, and the creation of isomorphic buffer zones around a feature).
Whilst these are, in many cases, a prerequisite to spatial *analysis*, they will
not be taken to constitute *analysis* for the purposes of this chapter. For the
same reason, some applications of statistical techniques in GIS to address existing deficiencies in data selection and aggregation algorithms, such as areal
interpolation (Flowerdew and Green, 1991), error propagation (Heuvelink
et al., 1989; Arbia, 1989; Carver, 1991) and missing value interpolation (Krug
and Martin, 1990) are also precluded from the discussion, although they are
ultimately fundamental and important to analysis.

It might be thought that the above restrictions on the discussion should
result in a fairly well defined and well understood set of methods for subsequent consideration. Unfortunately, that is not the case. Statistical spatial
analysis encompasses an expanding range of methods which address many
different spatial problems, from image enhancement and pattern recognition,
through to the interpolation of sampled mineral deposits, the investigation
of spatial or spatio-temporal clustering of disease, the modelling of socio-
economic trends, and the study of human and animal migration. Many such
techniques were originally developed outside of the field of statistics; for
example in geography, geostatistics, econometrics, epidemiology, or urban
and regional planning and the wide range of relevant literature reflects this
(*Journal of Regional Science, Biometrics, Biometrika, Environment and
Planning (A), Geographical Analysis, Journal of Ecology, Journal of the
American Statistical Association, Journal of the Royal Statistical Society,
Series B., Applied Statistics, Journal of Soil Science* and contributions in many
other journals). This has resulted in some confusion in terminology and a fair
amount of reinventing the wheel (or at least something which is pretty round!)
and has led Goodchild (1991) to refer to the field as: 'a set of techniques
developed in a variety of disciplines without any clear system of codification
or strong conceptual or theoretical framework'.

However, several recent texts have brought such techniques together,
under a structured and unified framework clarifying links between hitherto
separate methodologies (Ripley, 1981; Diggle, 1983; Upton and Fingleton,
1985; Ripley, 1988; Anselin, 1988a; Upton and Fingleton, 1989; Haining, 1990;
Cressie, 1991). Naturally, each of these texts has had different emphases and
has focused on particular aspects of the field, but what emerges is something
that is clearly identifiable as *spatial statistics*. Not all such techniques are of
sufficiently wide application, or tractability, to be candidates for linkage or
integration into a GIS environment, but it is felt that they do provide the best
basis from which to start. It should be acknowledged that Openshaw (1991)
has argued strongly against this. He advocates the development of new, generic

spatial analysis methods, customised for a data-rich GIS environment, rather than for the linkage of traditional spatial statistical methods to GIS – the value of many of which he doubts in a GIS context. Whilst agreeing with his general points concerning the somewhat naïve spatial models implicit in some existing methods, and the challenge to some of their inherent statistical assumptions presented by the kind of data typical in a GIS context, this chapter takes the view that the kind of automated, computationally intensive, pattern searches that he has suggested (Openshaw *et al.*, 1987; Openshaw *et al.*, 1990) are firstly, not without their own set of problems (not least that they require computing facilities which make them infeasible to implement for the majority of GIS users at the current time) and secondly, need not necessarily be viewed as substitutes for, but rather as additions to, existing methods. It is therefore felt justified in limiting most of the subsequent discussion to links between GIS and better understood spatial statistics techniques, drawn from the existing literature; rather than consider more advanced 'customised' techniques, which are still the subject of research and are not yet developed to the stage where they are widely applicable.

The objective is therefore to identify a set of existing statistical spatial analysis techniques for which it may be valuable to discuss closer links to GIS. Those techniques selected should cover the scope of analytical problems which most commonly arise; be sophisticated enough to benefit from the more realistic representation of space which GIS can provide; be of fairly wide application across disciplines; be computationally feasible; cater for informal graphical, exploration of spatial heterogeneity; and finally, fall into a relatively small number of core groups which are conceptually related and may therefore be considered together in respect of links to GIS. The methods chosen are presented in Table 2.1.

The division in Table 2.1 by data structure is common to most discussions of spatial statistics. *Locational data* consists purely of the locations at which a set of events occurred. This is often referred to as *event data*, *object data* or a *point process*. A typical example might be the locations of cases of some disease within a study area. The multivariate case arises where different types of events are involved – the *marked point process*. A temporal aspect to such data is often present and can be treated as a special kind of *marking*. *Attribute data* consists of values, or attributes, associated with a set of locations; in general the latter may be specific points, cells of a regular grid or irregular polygons. Typical examples might be soil property at sampled point locations in a field, a remote sensing measurement on a regular grid, or a mortality rate within irregular census tracts. A distinction is often made between the *point*, *regular grid or raster* and *irregular polygonal* case, because certain variants of analysis may not be relevant in all cases, but, conceptually, these all belong to the same data structure and similar models apply – interest is in analysing spatial variation in attribute values, *conditional* on the locations. The multivariate case arises when a vector of attributes is present at each location, one of which may be a temporal element. Finally, *interaction data* consists of

Table 2.1. *Potentially useful statistical spatial analysis techniques in relation to GIS.*

Data Structure	Dimensionality	
	Univariate	**Multivariate**
Locational Data	Nearest Neighbour Methods K-Functions	
		Bivariate K Functions Space Time Interaction
	Kernel Density Estimation Kernel Regression	Adaptive Kernel Density Estimation Kernel Regression
	Bayesian Smoothing – ICM	Bayesian smoothing – ICM
Attribute Data	Spatial Autocorrelation Spatial Correlograms Variograms	Multivariate Spatial Correlation
	Trend Surface Analysis Kriging	Spatial Regression Co-Kriging
		Spatio-temporal models
		Spatial General Linear Modelling
		Cluster Analysis Canonical Correlation Multidimensional scaling
Interaction Data	Spatial Interaction Methods	Augmented Spatial Interaction models

quantitative measurements each of which is associated with a link, or pair of locations; these are normally two points, but this could be generalized to mixtures of points and regular or irregular areas. A typical example might be flows of individuals from places of residence to retail shopping outlets. In the multivariate case such data may be supplemented by a vector of measurements at link origins which characterize *demand* and at link destinations which characterize *attractiveness*.

The division by dimensionality in Table 2.1 is felt useful to emphasize the difference between the case where only a single pattern is being investigated (univariate), and that where more than one pattern is involved (multivariate); in the case of the latter one may wish to study relationships between patterns or attempt to 'explain' one pattern of particular interest in terms of the others. The essential difference is that in the univariate case the only possible 'explanation' of observed pattern is in terms of some overall spatial 'trend' across the locations, or possibly through a tendency for neighbours to 'move together', i.e. to be 'spatially autocorrelated'; whereas in the multivariate case additional 'explanation' is available via other 'spatially correlated' attributes measured at each of the locations.

This chapter is not the place for a detailed technical discussion of the techniques presented in Table 2.1, but references and brief overviews will be given in the next section, when the potential benefits of links between such techniques and GIS are discussed. It will also be useful then to discuss techniques under some general sub-headings and the boxing in Table 2.1 attempts to pre-empt this by indicating techniques which are conceptually related, either within data structure, or across dimensionality.

Before proceeding to that discussion, it may be valuable to make some general points concerning the techniques presented. Firstly, these are fairly sophisticated methods and would all require the existence of 'lower level' functionality (such as simple statistical graphics and the ability to transform data and compute basic summary statistics) which has not been explicitly included in Table 2.1. This point will be followed up in more detail in the next section. On a related theme, the question of whether some of the standard non-spatial statistical techniques such as ordinary regression, analysis of variance or log linear modelling should be available to analyse spatial data, does not arise given the techniques included in Table 2.1. The majority of non-spatial analyses arise as special cases of the techniques that are included and would therefore be implicitly available.

Secondly, the classification given does not imply that some types of data cannot be analysed under more than one heading. For example locational data can be aggregated to a polygon coverage and then analysed using techniques for area data, although biases associated with the *modifiable areal unit problem* (Openshaw, 1984) may well make this undesirable. Similarly, attribute data aggregated over irregular polygons may also be interpolated to a finer, regular, or near regular, lattice. This may be particularly valuable in the case where the original polygon coverage relates to zones which are purely administrative and convey little of spatial relevance (Martin, 1991; Bracken, 1991). It may also be desirable to analyse each part of a multivariate data set by univariate methods, particularly at an exploratory stage. Univariate analyses would also of course be relevant to the analysis of residuals from a multivariate model.

Thirdly, no explicit distinction has been made between 'exploratory' or 'descriptive' techniques on the one hand, and 'confirmatory' or 'formalized modelling' techniques on the other. This reflects a belief that ultimately analysis involves a close, interactive, iteration between exploration of data and the identification and estimation of models. 'Exploratory' techniques may prove just as useful in analysing model validity as they do in suggesting the model in the first place. It is felt that an explicit distinction is unnecessary, what is more important is that a variety of tools are available, some of which allow minimal preconceptions as to the homogeneity or correlation structure present in the data. For this reason various nonparametric smoothing methods are felt to be essential and these have been included. At the same time various 'robust' parameter estimation techniques would be preferable options to include alongside their classical 'least squares' counterparts in relation to the spatial modelling techniques that have been suggested.

Fourthly, it could be argued that any statistical spatial analysis methods implemented in GIS should be robust and easy to understand, since they will be used in the main by nonspecialists. Some of the methods that have been suggested are fairly sophisticated and are certainly not without theoretical problems. For example, edge correction techniques are a difficult theoretical area in both K-functions and kernel smoothing techniques. Kernel smoothing over irregular areal units provides difficulties, as do estimation techniques for spatial general linear models for counts or proportions. Many questions also remain to be resolved in respect of spatio-temporal modelling for both locational and area data. However, the relative sophistication of such methods needs to viewed in the light of that required to carry out valuable analysis. Spatial analysis is conceptually a complex area; GIS increases this complexity by making more realistic assumptions about the study area available. It does not make sense to expect to be able to reduce the analysis tools necessary to cover the range of practical problems to a 'simple generic set'. In spatial analysis there is certainly some truth in the adage that the provision of methods which 'even a fool can use' will ensure that 'only a fool will find them useful'. The theoretical problems which remain are ultimately best resolved in an on-going practical context – spatial data analysis and GIS is a two-way process.

Fifthly, it is felt that the techniques presented are of sufficiently wide application for them to be generally useful. Clearly the emphasis and importance of techniques for different disciplines will tend to vary across the different rows and columns of Table 2.1, but none of these techniques are specifically of use in one discipline. Kriging (Isaaks and Srivastava, 1989) may historically have been almost exclusively of interest to geostatisticians, but the identification of kriging models through the variogram is closely related to the general identification of covariance structure which is involved in spatial econometric modelling, where virtually equivalent models are employed (Ripley, 1981). As a general method of statistical interpolation, kriging has much wider potential application than in the field of geostatistics. Similarly kernel density estimation (Silverman, 1986), which provides a spatially smooth estimate of the local intensity of events; may prove just as useful to the forester or the researcher studying patterns of crime, as to the epidemiologist.

A final point is that no suggestion is implied that one would consider it necessary, or even desirable, to fully integrate all the techniques in Table 2.1 into a GIS environment. Some of the techniques involve computationally intensive algorithms where numerical stability is important, some require sophisticated simulation and spatial resampling ability. Vendors of commercial GIS are unlikely to be persuaded that there is sufficient demand from users to merit the development of such facilities, given that the major markets for such products are in the management of large spatial databases within the utilities and local government, rather than in scientific research; nor are they likely to have the mathematical and statistical expertise required. Besides which, flexible, statistical computing languages such as S or S-Plus (Becker *et al.*, 1988) are already available to perform much of the algorithmic work, and it would seem unnecessary to duplicate this again within a GIS. The objective in this section was to identify a set of existing statistical spatial analysis techniques for which it may be valuable to discuss *closer links to* GIS, the argument being that it is necessary to appreciate what tools are useful from an analytical perspective, before considering what benefits might arise from the interaction of GIS with these tools, and what sort of linkage that kind of interaction would require. This latter issue is now taken up in more detail.

Potential benefits and progress in linking statistical spatial analysis and GIS

The techniques discussed in the previous section were identified as potentially useful statistical techniques to discuss in relation to GIS and divided into a number of conceptually related groups. The questions addressed in this section are '*what are the potential benefits of linking each group of these techniques to GIS?*' and '*how much progress has been made in achieving those potential benefits in each case?*'

The starting point for such a discussion must be to identify the general benefits of close links between GIS and statistical spatial analysis. In a sense each of the techniques listed in Table 2.1 may be thought of as an algorithm with specified inputs and outputs. The value of GIS to the spatial analyst is in enhancing either the quality of inputs, or the analysis of outputs, or both. It is suggested that all such benefits essentially fall under three general headings: (1) *flexible ability to geographically visualise both raw and derived data*, (2) *provision of flexible spatial functions for editing, transformation, aggregation and selection of both raw and derived data, and* (3) *easy access to the spatial relationships between entities in the study area.*

To take a specific example, suppose that data is available relating to occurrences of a particular type of event within some geographical area. The first stage in analysis might be to derive a spatially smooth kernel estimate of the intensity of occurrence, for which *straight line distances* between occurrences and the *spatial configuration of the study region* would be required. The results are then *visualized in conjunction with selected aspects of the geography and topography of the study area*. Following this one might want to *select regions in the study area either directly or via a spatial query based on various attributes of the areas*, and *zoom in on one or more such areas*. K-functions are now computed for these new areas, requiring as input *boundary configurations and more realistic distances, based perhaps on travelling time along a road network*. These are then *visualized in conjunction with dot maps of events within the appropriate region.*

The italics here attempt to indicate at which point GIS facilities become of use to the analyst, either in visualization of results, or in the provision of inputs which involve spatial selection, or the derivation of spatial relationships in the region under study.

It is strongly felt that for it to be worth the effort to link any spatial analysis method to a GIS there must be a significant payback in one of these three areas. Furthermore, the extent to which it is worthwhile pursuing a link for any particular method is determined by the degree of payback that is possible over the three areas. Goodchild (1991) has characterized the general types of links between GIS and spatial analysis as: *fully integrated, tightly coupled,* or *loosely coupled.* Expressed in these terms, the argument above implies that the type of coupling appropriate for any technique would be the one that maximized the potential payback inherent in that technique, either for the geographical visualization of results, or for transfer to the algorithm of GIS derived spatial selection results, or for transfer of GIS derived spatial relationships. There would be little point in integrating an analysis technique into GIS whose output was not particularly amenable to visual display in the form of a map, or useful in conjunction with a map, or which could not exploit the more sophisticated representations of spatial geography and topography that the GIS could supply.

Specific details are now given as to what each of these three areas of potential benefit in linking GIS to statistical spatial analysis might mean in respect of each of the groups of techniques presented in Table 2.1. The progress that has been made in achieving those specific potential benefits is then reported. Where necessary to clarify the discussion, a brief, non-technical introduction to techniques within the group, with relevant references, is also given.

Simple descriptive analyses, data transformation and summarization

It is logical to start with what might be termed the *hidden agenda* of lower level statistical techniques which are implicitly required by the type of functions listed in Table 2.1, but are not explicitly included there. These were briefly discussed in the previous section, and consist of simple statistical, graphical and numerical methods for summarising and manipulating data, (including histograms, scatter plots, box plots, simple summary statistics and data transformation). Such basic descriptive methods need no introduction, but the potential benefits to be realised in linking these to GIS should not be ignored, particularly as all other analytical methods ultimately depend on such elementary functions.

Potential benefits

In terms of the general framework presented earlier, many of the potential benefits in linking such elementary functions to GIS are in the area of geographical visualization. There is much value in being able to view simple statistical summaries and plots of data at the same time as being able to view the data geographically in the form of a map. Recent developments in statistical graphics (Cleveland and McGill, 1988), all acknowledge the value of being able to simultaneously window various different views of the data, some graphical, others purely numerical. This is particularly valuable if the various views can be cross-referenced by dynamically linking the windows. A user can interact with any of the views of the data to select or highlight particular points of interest and these are then automatically highlighted in any other appropriate views of the data. In the GIS case this would, for example, allow the user to select outlying points in a scatter plot; these would immediately then be highlighted in the map view.

Another area of potential benefits involves direct linkage of basic statistical techniques to the spatial query and selection functions of the GIS, and at the same time to derived spatial properties or relationships, such as area, distance, adjacency, or distance along a network. This would allow data transformation to draw on derived spatial aspects of the data as well as the range of normal mathematical and statistical functions. One could, for example, compute spatial averages of zones with a common boundary, or plot the correspondence between the values at locations and those at neighbouring locations, where neighbours could be defined by a number of different spatial criteria. Spatial query would allow the user to be able to interactively redefine the region studied and then summarize or plot values within that area; or, alternatively, to partition the map into different regions and compare various summaries or plots between them. These would provide particularly useful tools for the exploration of heterogeneity and in distinguishing between local and global properties of any spatial variation.

The combination of dynamic windowing with this sort of spatial selection offers even more possibilities. For example Diggle (in Haining and Wise, 1991) has suggested the possibility of being able to 'drag' a selection window through the map, with any associated statistical summary or graphical windows being constantly updated to reflect data values in the moving selection window.

Progress

Since most of the initiatives in linking or incorporating statistical functions into GIS have involved some techniques which can be described as basic descriptive or graphical methods, reference is necessarily made here to most of the initiatives which subsequently also arise in relation to the other, more specific, groups of statistical techniques which have been defined. This is convenient since it allows most of the developments to be introduced together and then briefly cross-referenced under each of the other headings as appropriate.

The progress that has been made in terms of linking basic statistical summarization and simple statistical graphics to GIS has been surprisingly slow. Some of the most commonly used large commercial GIS systems such as ARC/INFO, SPANS, or GENAMAP offer little support for basic statistical summarization, data transformation, or simple graphics such as scatter plots and frequency distributions, although clearly they must already contain much of the functionality required for this. Most of them do provide macro languages and 'hooks' for user developed functions written in a low level language, such as C or FORTRAN. But, in general, researchers have not found these to be conducive to developing statistical functions, because they offer no or few high level facilities for accessing the GIS data files or for statistical graphical display. One example of the possibility of using these alone is provided by Ding and Fotheringham (1991) who have developed a statistical analysis module (SAM) which consists of a set of C programs running entirely within ARC/INFO and accessed via the ARC macro language (AML). At a different level, one well known grid-based GIS system for the IBM PC, IDRISI (Eastman, 1990), offers a modular structure and an inherently very simple data structure which has successfully encouraged users to develop their own IDRISI modules for several forms of spatial analysis. A similar kind of initiative, on a somewhat grander scale, is represented by the GRASS system, which is the principal GIS used by the US National Parks and Soil Conservation Service, and provides raster and vector GIS routines, developed for UNIX machines, and available in the public domain, to which users can interface their own C routines. Interfaces are also available to a number of common DBMS systems and image processing packages. Links are also possible to the highly flexible and powerful statistical programming language S-Plus (Becker et al., 1988), which provides excellent facilities for dynamic, windowed, graphics.

Various progress has been reported in terms of loosely coupling a GIS to external statistical packages or graphics software. For example Walker and Moore (1988) discuss such a modular computing environment, with MINITAB, GLIM and CART built onto a central GIS core. Various other examples of links with MINITAB, SPSS and SAS have been reported. All of these have involved linkage through ASCII files exported from the GIS. Waugh (1986) has described a programmable tool to covert ASCII files from one format to another (Geolink), and which facilitates file transfer in such arrangements. Closer links have been achieved by Rowlingson et al. (1991), who have interfaced ARC/INFO to the graphics package UNIRAS in a relatively seamless way through specially developed FORTRAN procedures. Work by Kehris (1990a) has demonstrated the possibility of a similar kind of link between ARC/INFO and GLIM.

An alternative approach to calling statistical packages from GIS has been to add limited GIS features to existing statistical products. For example, SPSS and other statistical packages now have simple mapping facilities; however these provide no real spatial functionality. Potentially of much more interest is the work that has been reported in adding spatial functionality to S-Plus. For example, Rowlingson and Diggle (1991a) have developed a collection of S-plus spatial functions (SPLANCS) which allow an interactive interface between visual views of a point coverage and statistical operations. Other researchers have also developed S-Plus functions for various forms of spatial analysis. (e.g. Ripley at Oxford, or Griffith (in Haining and Wise, 1991) who refers to another collection known as GS+).

Another area of progress has been the development of free-standing packages which attempt to combine some limited GIS functionality with statistical analysis. The majority of such products have been developed for the IBM PC market. Mapping packages such as MAPINFO, ATLAS*GRAPHICS, MAPIS, and GIMMS all offer various forms of basic descriptive or non-spatial statistical analyses combined with choropleth or dot mapping, but little ability to interact with the spatial properties of such maps. INFO-MAP (Bailey, 1990a; Bailey, 1990b) offers a language for data transformation which contains a number of spatial functions such as straight-line distance, area, perimeter, adjacency or nearest neighbours of different orders, and then allows summarization or plots of the results to be windowed onto the displayed map. However, such windows are not dynamically linked to the map, and the package has no real GIS functionality in terms of topographic detail, spatial query or map overlay. More effective use of the potential of GIS functionality has been achieved by Haslett *et al.* (1990), who have exploited the Apple Macintosh environment to develop SPIDER (now called REGARD), a powerful package which offers dynamically linked views of the data, combined with a language which includes spatial functions. They demonstrate, for example, how one can highlight points of interest on scatter plots or variogram clouds in one window and immediately observe where these contributions arise geographically in the map view. The package also permits several layers of data to be associated with a map and allows calculations to be carried out between these layers. It is felt that SPIDER begins to come close to exploiting the real potential of linking statistical tools to GIS functionality.

Nearest neighbour methods and K-functions

Moving on to the first of the groups of methods explicitly listed in Table 2.1, nearest neighbour methods involve a method of exploring pattern in locational data by comparing graphically the observed distribution functions of event-to-event or random point-to-event nearest neighbour distances, either with each other or with those that may be theoretically expected from various hypothesized models, in particular that of spatial randomness (Upton, 1985). The K-function (Ripley, 1977) looks at all inter-event distances rather than just those to nearest neighbours, $K(d)$ being defined as the expected number of further events within a distance d of an arbitrary event. Graphical or statistical comparison of the observed K-function with those simulated from possible explanatory models over the same study area allows assessment as to whether the observed occurrences are likely to have arisen from any particular one of these models. In the multivariate case of a 'marked' point process bivariate K-functions can be used in a similar way (Lotwick and Silverman, 1982). Extensions to these sort of techniques are available to deal with situations involving spatio-temporal patterns.

Potential benefits

The potential benefits involved in linking these sort of methods to GIS are largely related to exploiting GIS functions for interactive deletion and insertion of events from the data set by selecting directly from the map; or for the dynamic definition of the study area either directly or by spatial search; or for regionalizing the study area by grouping together homogeneous sub-areas and then using point-in-polygon functions to identify events in each of these areas. There are also potential benefits in the use of area and perimeter calculations, in the use of more realistic inter-event connectivity measures, derived directly from the topology of the area, and in identifying neighbours of different order on the basis of various connectivity measures, for example road network distances might be more relevant than straight line distance in the case where the events are traffic accidents. Other benefits of GIS functionality relate to a general problem with these kinds of analyses which involves correcting for the distortion that the edge of the study region will introduce into observed inter-event distances. Correction techniques (Rowlingson and Diggle, 1991b) require detailed knowledge of the shape of the boundary. GIS functions would be particularly useful in this regard and introduce a number of new possibilities such as investigating whether different edge corrections should be applied to boundaries which are qualitatively different, for example coast-lines as opposed to administrative zones.

The potential benefits arising from geographical visualization of the results of these methods is less clear. Certainly it is useful to be able to window K-functions whilst viewing the underlying study area, and Diggle (in Haining and Wise, 1991) has suggested a situation where the K-function might be updated dynamically as one moved a selection window around the area. Dynamic visualisation of spatio-temporal patterns (Charlton et al., 1989) – watching the process develop in space and time – is another potentially valuable exploratory tool.

Progress

There are several examples of exploiting GIS functionality in this general area of statistical methods. Openshaw's GAM (Openshaw et al., 1987) is a stand-alone package, developed specifically to investigate the clustering of rare disease. At the PC level, the IDRISI system mentioned earlier provides some modules for simple descriptions of point patterns, but only at the level of aggregated quadrat counts. INFO-MAP, also mentioned above, provides for basic analyses of nearest neighbour event–event distances and, in the most recent version, for the calculation and display of univariate K-functions. SPIDER, the Apple Macintosh system also referenced above, more effectively exploits visualization benefits by allowing for dynamic spatial selection of the study area, combined with a large range of possibilities for the exploratory analysis of interevent distances through its versatile analysis language. Rowlingson *et al.* (1991) have developed FORTRAN code, called from ARC/INFO, which accesses a point coverage and computes a univariate K-function, but currently this does not access the boundary of the study region directly; it assumes a fixed polygonal study area. More flexibility has been achieved in SPLANCS, the spatial analysis routines in S-Plus (Rowlingson and Diggle, 1991a) mentioned above, which allows the calculation of univariate and bivariate K-functions for imported point coverages together with an imported boundary polygon. Simulation of the equivalent functions under random dispersion of events within the same boundary is also provided. These S-Plus functions make little use of GIS functionality; all spatial structures such as area and edge corrections are rebuilt in S-Plus. In the specific area of space-time analyses, Charlton *et al.* (1989) report some work in using dynamic visualization through what they refers to as *spatial video modelling*.

Kernel and Bayesian smoothing methods

Statistical smoothing consists of a range of nonparametric techniques for filtering out variability in a data set whilst retaining the essential local features of the data. In a spatial context these may be particularly valuable exploratory techniques for identifying *hot spots* or areas of homogeneity, for identifying possible models and for analysing how well models fit the observed data.

Various simple smoothing ideas are available, such as spatial moving averages, or, for regular grids, median polish (Cressie, 1984; Cressie and Read, 1989). However, a more sophisticated and general class of such methods stems from the idea of kernel smoothing (Silverman, 1986). Here the smoothed value at any point is essentially estimated by a weighted average of all other values, with the weights arising from a probability distribution centred at that point and referred to as the kernel. The degree of smoothing is controlled through choice of a parameter known as the band width, which may be set to reflect the geographical scale of some particular hypothesis of interest, or optimally estimated as part of the smoothing process by cross-validation techniques. Kernel density estimation (Diggle, 1985) relates to locational data and refers to a kernel method for obtaining a spatially smooth estimate of the local intensity of events over a study area, which essentially amounts to a 'risk surface' for the occurrence of those events. Adaptive Kernel density estimation (Silverman, 1986) is an extension where the band-width parameter is automatically varied throughout the region to account for the effect of some other possible related measure, such as population at risk. Kernel regression (Silverman, 1986) relates to the situation where smoothing is required for an attribute or quantitative response which has been measured over a set of locations, rather than the occurrence of discrete events, and so would be an appropriate tool for area data.

Another approach to smoothing arises from research in image processing into how to 'reconstruct' a scene from a 'messy' image, using the knowledge that pixels close together tend to have the same or similar colours (Besag, 1986). This may be particularly useful in the case of area data which has been collected on a fine regular grid or which can be decomposed to that structure. A prior assumption that the local characteristics of the true scene can be represented by a particular type of probability distribution known as a Markov random field is combined with the observed data using Bayes' theorem and the true scene estimated in order to maximize a-posteriori probability. Besag (1986) has suggested a computationally efficient method which he refers to as 'Iterated Conditional Modes' or ICM.

Potential benefits

The potential benefits of linking such techniques to GIS are mainly in the area of being able to visualise the resulting surfaces as contours or 3-D projections, in relation to underlying geographical and topographical features. They may also be of use in the identification of local covariance structures. One could for example envisage a spatially smoothed map of a variogram. There are also potential benefits in being able to combine smoothing with spatial query. Automatic smoothing may be meaningless in the context of physical geography. For example, smoothing that crossed certain physical features might be undesirable. Rather than attempt to build such structures into the algorithm, which may be very complex, it might be a better solution to allow a user to be able to interactively define the regions over which smoothing is to occur. Potential benefits in linkage to GIS in other areas is less convincing; there may be some benefits to exploiting GIS functionality in correcting for edge effects in smoothing and in the derivation of spatial connectivity and adjacency, particularly in the smoothing of irregular attribute data – an area in which these techniques are relatively undeveloped.

Progress

SPIDER, mentioned previously, allows the user to create a region, for which a moving average can be computed, manually moving this region around the map thus, the user generates new views of the data, this idea is effectively computing smoothed statistics over the map. Simple spatial moving average smoothing is also available in INFO-MAP. However, currently no GIS system directly supports either kernel or Bayesian smoothing algorithms, although they are implemented in some image processing packages. Brunsdon (1991) provides an example of exporting data from a GIS, applying adaptive kernel density estimation and then importing the resulting surface for display and use in further analysis. Rowlingson *et al.* (1991) describe the implementation of a FORTRAN module in ARC/INFO for the modelling of the raised incidence of events around a point source, which employs kernel density estimation. Currently the optimal band-width needs to estimated by a separate program which runs independently of the GIS. The S-plus spatial functions of Rowlingson and Diggle (1991) provide for optimal kernel density estimation with edge corrections (Berman and Diggle, 1989) for a given set of points within a defined polygon. No links with GIS have been reported in respect of the other smoothing methods under this general heading. Although it is perhaps worth remarking here on the possibilities for using neural networks in GIS smoothing algorithms as opposed to the statistical algorithms that have been discussed, Openshaw *et al.* (1991) have discussed the use of neural networks in GIS in a more general context.

Spatial autocorrelation and covariance structure

This class of related techniques are all concerned with exploring spatial covariance structure in attribute data, i.e. whether, and in what way, adjacent or neighbouring values tend to move together. In the univariate case they range from standard tests for autocorrelation, based on statistics such as Moran's I or the Langrange multiplier (Anselin, 1988b), to the derivation and interpretation of plots of the autocorrelation at different spatial separations or lags – known as autocorrelograms; or of plots of the mean squared difference between data values at different spatial separations along particular directions in the space – known as variograms (Isaaks and Srivastava, 1989). In the multivariate case, multivariate spatial correlation enables correlation to be assessed between two measurements allowing for the fact that either measure may itself be autocorrelated within the space. Getis (1990) has also developed a type of second order analysis, based on an extension of K-function concepts to attribute data, for describing the spatial association between weighted observations.

Such methods are of use in a general exploratory sense to summarize the overall existence of pattern in attribute data and to establish the validity of various stationarity assumptions prior to modelling. In particular, they are fundamental to identifying possible forms of spatial model for the data. They also provide important diagnostic tools for validating the results of fitting any particular model to the data. The comparison of observed variograms with those that might be expected from particular theoretical models is particularly used in geostatistics in relation to kriging (Isaaks and Srivastava, 1989), but variograms are also useful in a more general context to assess covariance structure.

Potential benefits

The potential benefits of linking these methods to GIS are largely in the area of facilitating the construction of proximity matrices between locations, often known as W matrices, which are a necessary input to many of the auto-correlation methods. These are often constructed in terms of Euclidean distance, adjacency, existence of a common boundary, or length of a shared perimeter, but the potential exists to derive more sophisticated relationship measures between areal units which account for physical barriers, involve network structures, or are perhaps based on additional interaction (flow) data. Spatial selection of regions also has applications particularly in enabling the easy study of correlation structures in particular sub-areas of the whole space which may differ markedly from the global picture. The ability to interactively define directions in the space for the calculation of variograms is also valuable in order to assess whether variation is *stationary* (purely a function of relative position; i.e. separation and direction) and if so whether it is *isotropic* (purely a function of separation and not direction) as opposed to *anisotropic* (a function of both separation and direction).

The potential benefits in terms of visualization of outputs in conjunction with maps is less convincing, unless dynamic links can be generated between contributions to the correlogram or variogram and areas in the map. The study of variogram clouds (Chauvet, 1982) – plots of the average squared difference between each pair of values against their separation – might provide a basis for this as well as the second order techniques described by Getis (1990).

Progress

There have been various attempts to link statistical methods under this heading to GIS. Some of the relevant work has already been mentioned under previous headings. The IDRISI system allows for the estimation of an autocorrelation coefficient at various spatial lags for grid-based data. The basis underlying the ARC/INFO link to GLIM, developed by Kehris, has also been used for the computation of autocorrelation statistics by direct access to the ARC/INFO data structures (Kehris, 1990b). The ARC/INFO spatial analysis module, SAM, developed by Ding and Fotheringham (1991) concentrates on the derivation of, measures of spatial autocorrelation and spatial association and makes direct use of the topological data structures within ARC/INFO to derive proximity measures. Lowell (1991), as part of a larger study, describes the use of the Software Tool Kit modules within the ERDAS GIS, to develop routines to compute spatial autocorrelation for a residual surface of continuous data, where the GIS is used directly to derive a polygon connectivity matrix. The PC mapping system INFO-MAP provides spatial autocorrelograms windowed onto the displayed map, but restricted to spatial lags defined in terms of successive nearest neighbours. SPIDER, developed by Haslett *et al.* (1990), allows the user to interactively define areas of the space and then produce dynamic plots of the values at locations against those at neighbours, within those areas. This provides a useful tool for informal investigation of the local covariance structure. The system also allows for the investigation of variogram clouds in a similar way. Anselin (1990a) has developed a stand-alone package using GAUSS for the IBM PC, called SpaceStat, which involves both exploratory analysis and formal tests for spatial auto-correlation and multivariate correlation. Openshaw *et al.* (1990) describe a specialized system GCEM to automatically explore digital map databases for possible correlations between various attribute coverages. Numerous studies have also been reported which involve exporting data from a GIS and then computing variograms or correlograms using special purpose software or statistical packages such as MINITAB, SPSS or SAS. Vincent and Gatrell (1991) provide a typical example, they use UNIMAP, a sub-package of the graphics package UNIRAS, to estimate experimental variograms and inter-actively fit a range of theoretical models to these, as part of a kriging study on the spatial variability of radon gas levels.

Geostatistical and spatial econometric modelling

This is a well developed set of methods for the univariate or multivariate modelling of attribute data, discussed at length in many of the standard texts on spatial statistics (e.g. Haining, 1990; Cressie, 1991). Essentially they consist of spatial extensions to the familiar family of standard elementary linear regression models for non-spatially related data. Spatial variation is modelled as consisting of a global trend in mean value together with a stationary local covariance structure, or propensity of values to 'move together'. In general the covariance structure is expressed in one of a number of simple parametric forms which, in practice, amounts to including various neighbouring (or spatially lagged) values of both response and explanatory variables into the regression model. Parameters are then estimated by maximum likelihood or generalized least squares. Other more robust estimation methods have also been developed (Anselin, 1990b). Extensions are available to deal with the case where a temporal component is also present.

The reader may be surprised to find kriging and co-kriging listed together with the above methods. These statistical interpolation techniques arise from the geostatistical literature (Matheron, 1971). They are based on local weighted averaging, appropriate weights being identified via an analysis of the vario-gram (Isaaks and Srivastava, 1989). Their value is that they avoid many of the somewhat unrealistic assumptions associated with traditional deterministic interpolation methods based on tessellation or trend surfaces (Oliver and Webster, 1990) and in addition, provide estimates of the errors that can be expected in the interpolated surface. Although the kriging approach appears to differ markedly from that of spatial regression modelling, kriging is essentially equivalent to prediction from a particular type of spatial regression (Ripley, 1981; Upton, 1985) and is therefore included with those techniques in this discussion.

Potential benefits

Some of the potential benefits that might arise by linking this general class of techniques closely to GIS, those which concern the derivation of more realistic spatial weight matrices and the exploration of spatial covariance structure, have already been discussed previously. In addition to these there are benefits which arise from being able to visualize results either in the form of a fitted surface or a residual map to assist in the identification of outliers or leverage points. This is particularly true in the case of kriging where simultaneous geographical display of the interpolated surface along with its associated prediction errors has considerable value. Indeed, there is an argument for kriging to be adopted as a basic method of surface interpolation in GIS as opposed to the standard deterministic tessellation techniques which currently prevail and which can produce artificially smooth surfaces. However, it has to be acknowledged that in the general case the numerical estimation of spatial regression models can be computationally intensive involving a large and possibly asymmetric matrix of spatial weights and is probably best dealt with by specialized software rather than integrated into GIS. This is also true of robust estimation methods based on bootstrap and/or jackknife resampling techniques.

Progress

Progress in linkage of GIS to analytical tools in this area of methods largely mirrors that in the area of autocorrelation and covariance structure. PC packages like IDRISI provide modules for regression (based on ordinary least squares), as does INFO-MAP. In the latter case use of nearest neighbour, location and distance functions in combination with o.l.s. regression allows the fitting of some simple forms of trend surface and spatial autoregressive models. The stand alone package SpaceStat developed by Anselin (1990a) includes estimation techniques for a wide range of spatial econometric models including spatio-temporal models. The ARC/INFO link to GLIM, developed by Kehris, can also be used for simple forms of regression modelling. As in the case of autocorrelation analyses, several studies have been reported which involve exporting data from a GIS and then fitting spatial regression models in statistical packages. For example, macros are available in MINITAB and SAS for some forms of such models (Griffith, 1988). There has been little work reported which has adopted robust estimation procedures (Anselin 1990b). Oliver and Webster (1990) discuss kriging in relation to GIS, in detail. Currently kriging is carried out by special purpose software outside the GIS. The example by Vincent and Gatrell (1991) quoted above used kriging options available in the UNIRAS KRIGPAK, but several other software packages exist, in some cases offering more advanced kriging options. The main point is of course that kriging done in this way makes no direct use of the GIS knowledge about the topology of the area, natural barriers, watersheds etc. which could be valuable in some applications. Kriging is an option in the new version 6 of ARC/INFO, but it remains to be seen how closely the option can be integrated with topological information.

Spatial general linear modelling

Spatial general linear models essentially extend the spatial regression models discussed in the previous section to cases where the attribute being modelled is purely categorical, or represents a count or a proportion and thus requires special consideration. They consist of spatial generalizations to the ideas of the log-linear modelling of contingency tables and the modelling of Poisson or binomial distributed variables (Aitkin *et al.*, 1989). The spatial forms of these models are relatively undeveloped, and involve a number of theoretical problems – standard statistical software for fitting such models, such as GLIM, is currently not able to deal with the non-diagonal weight matrices involved in the spatial case. However, they are included here to emphasise the need for special methods to handle the spatial modelling of attributes which are qualitative or consist of counts or proportions, since these commonly arise in spatial socio-economic data sets.

Potential benefits

Potential benefits of linking these methods to GIS are ultimately similar to those for spatial regression models. Currently, however they are considerably limited by the theoretical problems that remain to be resolved concerning the spatial forms of such models.

Progress

There has been almost no work reported which has involved using spatial general linear modelling in conjunction with GIS. The work by Kehris involving a link between ARC/INFO and GLIM would allow the fitting of non-spatial general linear models, and macros could be developed in GLIM to cope with some forms of spatial general linear model. For example Flowerdew and Green (1991) have used this link to develop and estimate models for cross areal interpolation between incompatible zonal systems.

Multivariate techniques

The methods of modelling multivariate data discussed under the previous two headings are concerned with modelling the relationship between one response variable of particular interest and others that may 'explain' its spatial variation. There are often situations where several possible response variables need to be dealt with simultaneously and this brings into consideration a further wide range of traditional multivariate statistical techniques. The majority of such techniques discussed in standard texts (Krzanowski, 1988) are not specifically oriented to spatially dependent data, but they may still be useful as data reduction tools and for identifying combinations of variables of possible interest for examination in a spatial context. One can envisage cluster analysis as being useful for identifying natural clusters of observations in data space which may then be examined for geographical pattern. Methods have also been suggested for incorporating spatial constraints into such classification methods (Oliver and Webster, 1989). Canonical correlation analysis would enable the search for combinations of response variables which were maximally spatially separated. Multidimensional scaling would enable one to search for a possible geometric configuration of observations in data space which could then be related to geographical configuration.

Potential benefits

It is felt that there is an important potential benefit in close linkage of such techniques to GIS but that this lies mostly in the area of being able to explore possible spatial pattern through geographical visualization of the results.

Progress

Little reference has been made to the use of multivariate statistical analysis in conjunction with GIS. An interesting example is reported by Lowell (1991), involving ERDAS and SPANS in conjunction with SAS, which used discriminant analysis to assist in the spatial modelling of ecological succession. At another level the IBM PC system INFO-MAP, discussed earlier, implements both single linkage and K-means cluster analysis which can involve derived spatial relationships such as location, distance or adjacency. Results of clustering can immediately be displayed as a choropleth map or used in subsequent analyses. But no spatially constrained clustering is available and no alternative forms of multivariate analysis are currently provided.

Spatial interaction models

The general problem addressed in spatial interaction studies is the modelling of a pattern of observed flows from a set of origins to a set of destinations in terms of some measures of demand at origins, of attractiveness at destinations, and of generalised distance or cost of travel between origins and destinations. The models conventionally used are the general class of gravity models, which were originally proposed on intuitive grounds, but have been theoretically justified as the solution to various optimization problems, such as the minimisation of total distance travelled or the maximisation of entropy (Erlander and Stewart, 1990; Wilson, 1970). Such models may be constrained to reproduce the total observed flows at either the origins, or the destinations, or at both. From the statistical point of view such models may be thought of as examples of general linear models with parameters estimated by maximum likelihood or iterative reweighted least squares (Baxter, 1985; Bennett and Haining, 1985). They should be distinguished from the purely deterministic location/allocation models which are incorporated into several commercial GIS which typically assume that flows will always be to the nearest available destination.

Potential benefits

The main benefit of close links between these sorts of methods and GIS is felt to be in the area of using GIS functions to derive improved 'distance' measures such as distance along a network, or travelling time, or reflection of physical barriers to travel. There may be potential in the geographical visualisation of results in terms of identifying outlying flows and examining the fit of the model within sub-regions to identify the possible importance of factors which have not been explicitly included in the model (Bailey, 1991).

Progress

In general the network analysis modules of commonly used GIS are quite well developed. They allow the calculation of shortest paths with the possibility of arcs being assigned flow impedances and barriers to mimic obstacles such as bridges, one-way systems etc. Researchers have also found it possible to develop their own routines for more sophisticated analysis such as the 'travelling salesman problem' (Vincent and Daly, 1990). Systems such as ARC/INFO, SPANS and Transcad also provide functionality for location/allocation modelling and again it has been possible to develop additional analytical routines. For example de Jong *et al.* (1991) describe one such application using Genamap, and Maguire *et al.* (1991) another using ARC/INFO. Such modelling is purely deterministic and oriented towards an optimization problem, rather than the description and modelling of interaction or flow data. In the latter area use of GIS has been more restricted. Van Est and Sliepen (1990) describe using the GIS, SALADIN, interfaced with a transport modelling package TRIPS, to calibrate gravity models, where links were derived directly from the GIS, and the GIS was used to aggregate and analyse relevant socio-economic data such as population or employment. Bailey (1991) describes an application using gravity models in conjunction with a mapping package. In both cases data was exported from the GIS to the modelling package and vice versa. Closer links between analysis and GIS in the area of the understanding of interaction issues, are discussed by Miller (1991), who describes the use of space-time prisms in GIS in connection with the modelling of accessibility.

Summary

It is somewhat difficult to summarize the necessarily wide ranging discussion in this chapter. Without doubt, one general conclusion that emerges is that statistical spatial analysis and GIS is an area in which there is considerable potential, interest and activity. Furthermore that statistical spatial analysis and modelling need not be considered a diverse and loosely connected set of techniques, with little inter-disciplinary consensus as to what constitutes a core of robust and practically useful methods. On the contrary, there exists a theoretically coherent and well understood core of techniques, which can be identified as generally useful across disciplines. In addition, there are considerable potential benefits to be realized in closer links between GIS and this core of established techniques. There is little basis to argue that such existing methods are fundamentally inappropriate tools for statistical analysis in a GIS context. Undoubtedly, the spatial models associated with some of these methods are relatively crude, their ability to deal with complex topology is limited, and theoretical problems remain to be resolved in several areas – such as how to deal effectively with edge effects or qualitative differences in types of boundary; but essentially the point remains that such methods have considerably more spatial sophistication than that which is currently being exploited. The problem is not that we do not have appropriate methods, but that we are not using them effectively in conjunction with GIS.

Figure 2.1 attempts to summarize the potential benefits arising from linking GIS more closely with the groups of statistical spatial methods which have formed the basis for the discussion throughout most of this chapter. At the same time, it provides a similar picture for the progress that has been made towards realising these benefits. The benefits are divided into the three general areas identified previously, i.e. geographical visualization of outputs from analysis, provision of improved inputs to analysis by virtue of exploiting GIS functions for spatial search and aggregation, and provision of improved inputs by using GIS to derive more realistic spatial properties and relationships.

Figure 2.1 is necessarily a somewhat crude and subjective assessment. It was obtained by assigning a score of high, medium or low to the potential for each group of statistical techniques, under each benefit heading. Progress in respect of each group of methods was then assessed as realizing a high, medium or low proportion of the corresponding potential. The results suggest that the groups of methods for which close linkage with GIS would give the greatest overall benefits, are that of simple descriptive statistics, and that which relates to the analysis of covariance structure, although the benefits are composed differently in each case. Other areas such as smoothing methods, K-functions, kriging and spatial regression, have nearly equivalent overall potential benefits. The only real area in which moderate progress has been achieved is that concerned with geographical visualization. Very little progress, except perhaps in

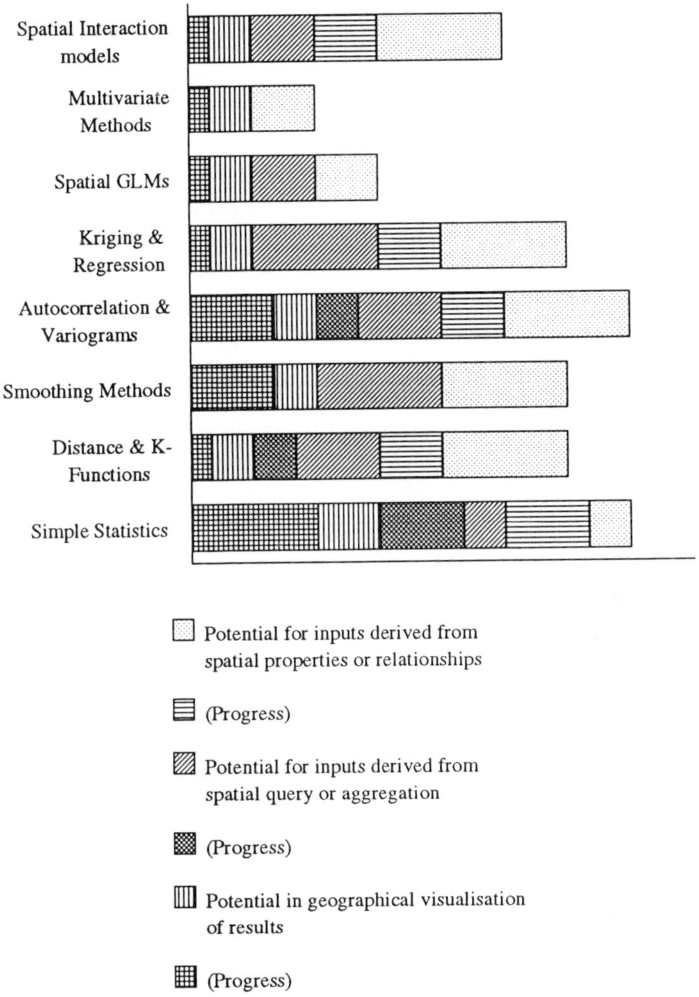

Figure 2.1. Summary of benefits in close linkage of GIS to different areas of statistical spatial analysis and progress towards realising those benefits.

simple descriptive statistical methods, has been achieved in the areas of benefit which relate to enhancing inputs to analysis by spatial search or the derivation of spatial relationships. The overall impression is that there remain considerable benefits yet to be realized in linkage with GIS, for all groups of methods. It should be borne in mind that the potential benefits represented are those relating to enhancing analytical methods by use of GIS functions – no attempt is being made here to address the overall usefulness of different kinds of analytical methods to the user community.

Conclusions

Turning to more general considerations raised by the discussion in this chapter, it is undoubtedly the case that for some time to come sophisticated forms of spatial analysis in conjunction with GIS are going to remain largely confined, as at present, to the research community. The reasons are firstly, that the largest users of GIS will continue to be organizations whose primary need is to manage large volumes of spatially-related data rather than to carry out sophisticated statistical analyses. Secondly, there is little expertise in sophisticated spatial statistics amongst GIS users and therefore a corresponding lack of pressure for such methods to be made available in conjunction with GIS. This situation is regrettable and could result in increasing volumes of spatially-referenced information being indiscriminately mapped, in the erroneous belief that mapping and analysis are in some sense the same thing. However, that aside, the implication is that fully integrated sophisticated, statistical techniques in large GIS are not likely to materialize in the short term.

At first sight this may also appear regrettable, but on reflection was probably never a practical proposition. The point has already been made that theoretical problems remain with some spatial analysis techniques. Also many of them involve simulation procedures rather than analytically exact results, which makes them difficult to implement in a sufficiently general form to suit a wide range of applications. The volume of data and extra topological detail, typically available from a GIS, compound such problems stretching current theory to the limit. Full integration demands a level of generality and robustness that it would currently be difficult to deliver in respect of many of the techniques. At the current time it is perhaps more valuable to retain the exploratory, graphical and computational flexibility of statistical environments like S-Plus rather than fossilize techniques. At the same time however, there is a need to exploit the potential benefits that GIS functionality can bring to spatial analysis techniques, and which have been demonstrated at length in this chapter.

The progress that has been made in developing flexible links with GIS through their embedded macro facilities, or through use of procedures written in low level languages has not been encouraging, as reported in previous sections. Undoubtedly, this will become a more efficient route as newer GIS software begins to pass on some of the benefits that object-oriented data structures will increasingly provide to those wishing to use these facilities. But realistically this route is only ever going to be of interest to a few enthusiastic specialized groups. The larger community of researchers wishing to use sophisticated statistical analysis in conjunction with GIS will need to adopt an open systems approach. The modern computing environment consists of workstations running GUI environments such as X-Windows communicating with various types of host via local area networks. This is even true at the PC or Apple Macintosh end of the market, where software to emulate

X-Windows terminals within a user's local GUI environment is becoming commonplace. A possible and attractive scenario for spatial analysis in conjunction with GIS is one of the GIS running in an X-Window to a remote host whilst the user has a variety of spatial analysis tools available locally in other windows. These tools could be general purpose statistical languages like S-Plus with added functions, which, as reported in this chapter, can cope with spatial data effectively. Or they could be stand alone spatial analysis systems which run within the GUI – self-contained enough to be a feasible development by research groups with particular analytical interests. SPIDER represents an excellent example of this kind of tool. The crucial question is then how to move data and spatial structures from the GIS window to the spatial analysis tool windows.

If one accepts this scenario, then this transfer problem is perhaps the key to linking sophisticated statistical spatial analysis and GIS. It is not an easy problem – the analyst has displayed a map in the GIS window, carefully created by spatial query and buffering as being the appropriate base for certain forms of analysis. However, many concepts which may be theoretically derivable from such a map are difficult to extract, such as connectivity, adjacency, travelling time, boundary configurations and presence of physical features such as rivers and coastline – what you can see is not necessarily what you can access. At the same time, it must be borne in mind that the kind of spatial analysis methods referenced in this chapter never require *all* that you can see, but rather a fairly simple abstraction of some aspects of it – all that you see is far too complicated to be practically usable at one time.

There is a close analogy here with large DBMS systems – modern PC spreadsheets like Excel, running in MS Windows, contain the functions to be able to dynamically issue SQL queries to DBMS systems such as ORACLE running on remote machines in another window. One challenge for the future in GIS is the development of an equivalent spatial SQL to allow users to access data which is displayed in a map window without the need to know about the particular data structures being used within the GIS. Although this is complex, the fact that it is only the displayed map window in which interest lies, may simplify the process. Recent developments in object oriented data structures for GIS may also help to make such developments easier. In the shorter term it may also be possible to proceed in simpler ways. The local spatial tools could attempt to rebuild spatial structures from fairly minimal information transferred from the map window (this is effectively the case with the S-Plus functions discussed in this chapter) or, perhaps from a bit-map pasted from that window, using raster to vector conversion.

If this is a sensible approach, then efforts should perhaps not be concentrated on integrating sophisticated spatial analysis into GIS; but rather, on developing this kind of interface and on developing a variety of the local spatial analysis tools to run alongside the map window. At the same time, there needs to be a greater effort in GIS to deal with those aspects of analysis which cannot be dealt with locally, including improving techniques for the derivation of proximity measures, for the estimation of missing values, for tracking error in spatial operations, or for interpolating data from incompatible zoning systems.

Acknowledgement

An earlier version of this chapter appeared in: *Proceedings of the 3rd European Conference on Geographical Information Systems*, Harts J., Ottens H., and Scholten H. (Eds.), EGIS Foundation, Faculty of Geographical Sciences, Utrecht, The Netherlands, Vol 1, 1992, 182–203.

References

Aitkin, M., Anderson, D., Francis, B. and Hinde, J. 1989, *Statistical Modelling in GLIM*, Oxford University Press.

Anselin, L. and Getis, A., 1991, Spatial statistical analysis and Geographical Information Systems, paper presented at *31st European Congress of the Regional Science Association*, Lisbon, Portugal.

Anselin, L., 1988a, *Spatial Econometrics, Methods and Models*, Kluwer Academic, Dordrecht.

Anselin, L., 1988b, Lagrange Multiplier Test Diagnostics for Spatial Dependence and Spatial Heterogeneity, *Geographical Analysis*, **20**, 1–17.

Anselin, L., 1990a, *SpaceStat: A program for the Statistical Analysis of Spatial Data*, Department of Geography, University of California, Santa Barbara.

Anselin, L., 1990b, Some robust approaches to testing and estimation in spatial Econometrics, *Regional Science and Urban Economics*, **20**, 141–163.

Arbia, G., 1989, Statistical effects of spatial data transformations: a proposed general framework, *Accuracy of spatial databases*, Goodchild, M. and Gopal, S. (Eds.), Taylor and Francis.

Bailey, T. C., 1990a, GIS and simple systems for visual, interactive spatial analysis, *The Cartographic Journal*, **27**, 79–84.

Bailey, T. C., 1990b, A Geographical Spreadsheet for Visual Interactive Spatial Analysis. *Proceedings of the first European Conference on Geographical Information Systems, Vol. 1*, Harts, J., Ottens, H. F. L. and Scholten, H. J. (Eds.), EGIS Foundation, Utrecht, The Netherlands, pp. 30–40.

Bailey, T. C., 1991, A Case Study employing GIS and Spatial Interaction Models in Location Planning. *Proceedings of the second European Conference on Geographical Information Systems, Vol. 1*, Harts, J., Ottens, H. F. L. and Scholten, H. J. (Eds.), EGIS Foundation, Utrecht, The Netherlands, pp. 55–65.

Baxter, M. J., 1986, Geographical and planning models for data on spatial flows, *The Statistician*, **35**, 191–198.

Becker, Chambers and Wilks, 1988, *The new S Language*, Pacific Grove: Wadworth and Brooks/Cole. California.

Bennett, R. J. and Haining, R. P., 1985, Spatial structure and spatial interaction models: modelling approaches to the statistical analysis of geographical data, *J. Royal Stat. Soc. (A)*, **148**, 1–27.

Berman, M. and Diggle, P. J., 1989, Estimating Weighted Integrals of the second order intensity of a spatial point process, *J. Royal Stat. Soc. (B)*, **51**, 81–92.

Besag, J. E., 1986, On the statistical analysis of Dirty Pictures, *J. Royal. Stat. Soc. (B)*, **48**, 259–279.

Bracken, I., 1991, A surface model of population for public resource allocation, *Mapping Awareness*, **5**, 35.

Brunsdon, C., 1991, Estimating probability surfaces in GIS: an adaptive technique, *Proceedings of the second European Conference on Geographical Information Systems*, Harts, J., Ottens, H. F. L. and Scholten, H. J. (Eds.), EGIS Foundation, Utrecht, Netherlands, pp. 155–163.

Burrough, P. A., 1990, Methods of Spatial Analysis in GIS, *Int. J Geographical Information Systems*, **4**, 221.

Carver, S., 1991, Adding Error Handling Functionality to the GIS Tool Kit, *Proceedings of the second European Conference on Geographical Information Systems*, Harts, J., Ottens, H. F. L. and Scholten, H. J. (Eds.), EGIS Foundation, Utrecht, Netherlands, pp. 187–194.

Cleveland, W. S. and McGill, M. E. (Eds.) (1988), *Dynamic Graphics for Statistics*, Pacific Grove, California: Wadsworth and Brooks/Cole.

Cressie, N. A. C., 1984, Towards Resistant Geostatistics, *Geostatistics for Natural Resources Characterisation, Part 1*, Verly, G., David, M., Journel, A. G. and Marechal, A. (Eds.), Dordrecht, Reidel, pp. 21–44.

Cressie, N. A. C. and Read, T. R. C., 1989, Spatial data analysis of regional counts, *Biometrical Journal*, **31**, 699–719.

Cressie, N. A. C., 1991, *Statistics for Spatial Data*, New York: John Wiley and Sons.

Charlton, M., Openshaw, S., Rainsbury, M. and Osland, C., 1989, *Spatial analysis by computer movie*, North East Regional Research Laboratory, Research Report 89/8, University of Newcastle.

Department of the Environment 1987, *Handling Geographical Information*. Report to Secretary of State for the Environment of the Committee of Enquiry into the Handling of Geographic Information Chaired by Lord Chorley, HMSO, London.

Diggle, P. J., 1983, *Statistical Analysis of Spatial Point Patterns*, London: Academic Press.

Diggle, P. J., 1985, A Kernel Method for smoothing point process data, *J. R. Stat. Soc. (C)*, **34**, 138–147.

Ding, Y. and Fotheringham, A. S., 1991, *The integration of spatial analysis and GIS*, working paper, NCGIA, Department of Geography, State University of New York, Buffalo.

Dixon, J., Openshaw, S. and Wymer, C., 1987, *A proposal and specification for a geographical subroutine library*, North East Regional Research Laboratory, research report RR87/3, University of Newcastle.

Eastman, J. R., 1990, *IDRISI: A Grid-Based Geographic Analysis System*, Department of Geography, Clark University, Worcester, Massachusetts.

Erlander, S. and Stewart, N. F., 1990, *The Gravity Model in Transportation Analysis – Theory and Extensions*, VSP, Utrecht, The Netherlands.

ESRI, 1990, The ARC/INFO Rev. 5.1 Release, *Mapping Awareness*, **4**, 57–63.

van Est, J. P. and Sliepen, C. M., 1990, Geographical Information Systems as a basis for Interaction Modelling, in *Proceedings of the first European Conference on Geographical Information Systems*, Harts, J., Ottens, H. F. L. and Scholten, H. J. (Eds.), EGIS Foundation, Utrecht, Netherlands, pp. 298–308.

Flowerdew, R. and Green, M., 1991, Data integration: statistical methods for transferring data between zonal systems, *Handling Geographical Information*, Masser, I. and Blakemore, M. (Eds.), Longman, London, pp. 18–37.

Getis, A., 1990, Screening for Spatial Dependence in Regression Analysis, *Papers of the Regional Science Association*, **69**, 69–81.

Goodchild, M. F., 1987, A spatial analytical perspective on Geographical Information Systems, *Int. J. Geographical Information Systems*, **1**, 335–354.

Goodchild, M. F., Haining, R. P. and Wise, S. M., 1992, Integrating GIS and Spatial Data Analysis: problems and possibilities, *Int. J. Geographical Information Systems*, **6**(5), 407–23.

Goodchild, M. F., 1991, Progress on the GIS research agenda, in *Proceedings of the second European Conference on Geographical Information Systems*, Harts, J., Ottens, H. F. L. and Scholten, H. J. (Eds.), EGIS Foundation, Utrecht, Netherlands, pp. 342–350.

Griffith, D. A., 1988, Estimating Spatial Autoregressive Model Parameters with Commercial Statistical Packages, *Geographical Analysis*, **20**, 176–186.

Haining, R. P. and Wise, S. M. (Eds.) (1991), *GIS and spatial data analysis: report on the Sheffield Workshop*, Regional Research Laboratory Initiative, Discussion Paper, 11, University of Sheffield.

Haining, R. P., 1990, *Spatial Data Analysis in the Social and Environmental Sciences*, Cambridge: Cambridge University Press.

Haslett, J., Wills, G. and Unwin, A., 1990, SPIDER – an interactive statistical tool for the analysis of spatially distributed data, *Int. J. Geographical Information Systems*, **4**, 285–296.

Heuvelink, G. B. M., Burrough, P. A. and Stein, A., 1989, Propagation of errors in spatial modelling with GIS, *Int. J. Geographical Information Systems*, **3**, 303–322.

Isaaks, E. H. and Srivastava, R. M. 1989, *An Introduction to Applied Geostatistics*, Oxford: Oxford University Press.

de Jong, T., Ritsema van Eck, J. R., Toppen, F., 1991, GIS as a Tool for Locating Service Centres, in *Proceedings of the second European Conference on Geographical Information Systems*, Harts, J., Ottens, H. F. L. and Scholten, H. J. (Eds.), EGIS Foundation, Utrecht, Netherlands, pp. 511–517.

Kehris, E., 1990a, *A geographical modelling environment built around ARC/INFO*, North West Regional Research Laboratory, Research Report 13, Lancaster University.

Kehris, E., 1990b, *Spatial autocorrelation statistics in ARC/INFO*, North West Regional Research Laboratory, Research Report 16, Lancaster University.

Krug, T. and Martin, R. J., 1990, *Efficient Methods for coping with missing information in remotely sensed data*, Research Report No. 371/90, Dept. of Probability and Statistics, University of Sheffield.

Krzanowski, W. J., 1988, *Principles of Multivariate Analysis*, Oxford: Clarendon Press.

Lowell, K., 1991, Utilising discriminant function analysis with a geographical information system to model ecological succession spatially, *Int. J. Geographical Information Systems*, **5**, 175–191.

Lotwick, H. W. and Silverman, B. W., 1982, Methods for analysing spatial processes of several types of points, *J. Royal Stat. Soc. (B)*, **39**, 172–212.

Maguire, D. J., Hickin, B., Longley, I. and V. Messev, T., 1991, Waste disposal site selection using raster and vector GIS, *Mapping Awareness*, **5**, 24–27.

Martin, D., 1991, Understanding Socio-economic Geography from the Analysis of Surface Form, *Proceedings of the second European Conference on Geographical Information Systems*, Harts, J., Ottens, H. F. L. and Scholten, H. J. (Eds.), EGIS Foundation, Utrecht, Netherlands, pp. 691–699.

Masser, I., 1988, The Regional Research Laboratory Initiative, *Int. J. Geographical Information Systems*, **2**, 11–22.

Matheron, G., 1971, The theory of regionalised variables and its applications, *Les Cahiers du Centre de Morphologie Mathematique de Fontainebleau*, No. 5, Paris.

Miller, H. J., 1991, Modelling accessibility using space-time prism concepts within Geographical Information Systems, *Int. J. Geographical Information Systems*, **5**, 287–301.

NCGIA, 1989, The research plan of the National Centre for Geographic Information and Analysis, *Int. J. Geographical Information Systems*, **3**, 117–136.

Oliver, M. A. and Webster, R., 1989, A geostatistical basis for spatial weighting in multivariate classification, *Mathematical Geology*, **21**, 15.

Oliver, M. A. and Webster, R., 1990, Kriging: a method of interpolation for geographical information systems. *Int. J. Geographical Information Systems*, **4**, 313–332.

Openshaw, S., 1984, The modifiable Areal Unit Problem, *Concepts and Techniques in Modern Geography 38*, Geo Books: Norwich.

Openshaw, S., 1991, A spatial analysis research agenda, *Handling Geographical Information*, Masser, I. and Blakemore, M. (Eds.), London: Longman.

Openshaw, S., Brunsdon, C. and Charlton, M., 1991, A spatial analysis tool kit for GIS, *Proceedings of the second European Conference on Geographical Information Systems*, Harts, J., Ottens, H. F. L. and Scholten, H. J. (Eds.), EGIS Foundation, Utrecht, Netherlands, 788–796.

Openshaw, S., Charlton, M., Wymer, C. and Craft, A., 1987, A Mark 1 Geographical Analysis Machine for the automated analysis of point data sets, *Int. J. Geographical Information Systems*, **1**, 335–358.

Openshaw, S., Cross, A. and Charlton, M., 1990, Building a prototype Geographical Correlates Exploration Machine, *Int. J. Geographical Information Systems*, **4**, 297–312.

Openshaw, S., 1990, Spatial analysis and geographical information systems: a review of progress and possibilities, *Geographical Information Systems for Urban and Regional Planning*, Scholten, H. J. and Stillwell, J. C. H. (Eds.), Dordrecht: Kluwer Academic Publishers, 153–163.

Rhind, D., 1988, A GIS research agenda, *Int. J. Geographical Information Systems*, **2**, 23–28.

Rhind, D. and Green, N. P. A., 1988, Design of a geographical information system for a heterogeneous scientific community, *Int. J. Geographical Information Systems*, **2**, 171–190.

Ripley, B. D., 1977, Modelling spatial patterns (with discussion), *J. Royal Stat. Soc. (B)*, **39**, 172–212.

Ripley, B. D., 1981, *Spatial Statistics*, John Wiley and Sons: New York.

Ripley, B. D., 1988, *Statistical Inference for Spatial Processes*, Cambridge: Cambridge University Press.

Rowlingson, B. S. and Diggle, P. J., 1991a, *SPLANCS: Spatial point pattern analysis code in S-Plus*, Mathematics Department Technical Report MA91/63, Lancaster University, Lancaster, UK.

Rowlingson, B. S. and Diggle, P. J., 1991b, *Estimating the K-function for a univariate point process on an arbitrary polygon*, Mathematics Department Technical Report, Lancaster University, Lancaster, UK.

Rowlingson, B. S., Flowerdew, R. and Gatrell, A., 1991, *Statistical Spatial Analysis in a Geographical Information Systems Framework*, North West Regional Research Laboratory, Research Report 23, Lancaster University.

Silverman, B. W., 1986, *Density Estimation for Statistics and Data Analysis*, London: Chapman and Hall.

Upton, G. J. and Fingleton, B., 1985, *Spatial Statistics by Example Vol. 1: Point pattern and Quantitative Data*, New York: John Wiley and Sons.

Upton, G. J. and Fingleton, B., 1989, *Spatial Statistics by Example Vol. 2: Categorical and Directional Data*, New York: John Wiley and Sons.

Vincent, P. and Daly, R., 1990, GIS and Large Travelling Salesman Problems, *Mapping Awareness*, **4**, 19–21.

Vincent, P. and Gatrell, A., 1991, The spatial distribution of radon gas in Lancashire (UK) – a kriging study, in *Proceedings, second European Conference on Geographical Information Systems*, Harts, J., Ottens, H. F. L. and Scholten, H. J. (Eds.), EGIS Foundation, Utrecht, Netherlands, pp. 1179–1186.

44   *T. C. Bailey*

Walker, P. A. and Moore, D. M., 1988, SIMPLE: An Inductive modelling and mapping tool for spatially oriented data, *Int. J. Geographical Information Systems*, **2**, 347–364.

Waugh, T. C., 1986, The Geolink system, interfacing large systems, *Proc. Auto Carto London Vol. 1*, Blakemore, M. J. (Ed.), London: RICS, pp. 76–85.

Wilson, A. G., 1970, *Entropy in Urban and Regional Modelling*, London: Pion.

Wise, S. and Haining, R., 1991, The role of spatial analysis in Geographical Information Systems, paper presented to the *Annual Conference of the Association of Geographic Information*, Birmingham.

3

Designing spatial data analysis modules
for geographical information systems

Robert Haining

Introduction

Spatial analysis (SA) is sometimes defined as a collection of techniques for
analysing geographical events where the results of analysis depend on the
spatial arrangement of the events (see, for example, Goodchild in Haining
and Wise (1991), Chorley (1972)). By the term 'geographical event' (hence-
forth, 'event') is meant a collection of point, line or area objects, located in
geographical space, attached to which are a set of (one or more) attribute
values. In contrast to other forms of analysis, therefore, SA requires information *both* on attribute values *and* the geographical locations of the objects
to which the collection of attributes are attached.

Based on the systematic collection of quantitative information, the aims of
SA are: (1) the careful and accurate description of events in geographical
space (including the description of pattern); (2) systematic exploration of the
pattern of events and the association between events in space in order to gain
a better understanding of the processes that might be responsible for the
observed distribution of events; (3) improving the ability to predict and control events occurring in geographical space.

Wise and Haining (1991) identify three main categories of SA which are
labelled statistical spatial data analysis (SDA), map based analysis and
mathematical modelling. This paper is concerned only with SDA. This is one
area of SA where there is widespread agreement on the benefits of closer
linkage with Geographic Information Systems (Haining and Wise, 1991). If
GIS is to reach the potential claimed by many of its definers and proponents
as a general purpose tool for handling spatial data then GIS needs to incorporate SDA techniques. If SDA techniques are to be of wider use to the
scientific community then GIS with its capability for data input, editing and
data display offers a potentially valuable platform. Without such a platform
and associated SDA software, the start up costs for rigorous analysis of spatially
referenced data can prove prohibitive.[1]

This chapter considers the first five of six questions that arise if linkage
between SDA and GIS is to take place.[2]

1. What types of data can be held in a GIS?
2. What classes of questions can be asked of such data?

3. What forms of SDA are available for tackling these questions?
4. What set of (individual) SDA tools is needed to support a coherent programme of SDA?
5. What are the fundamental operations needed to support these tools?
6. Can the existing functionality within GIS support these fundamental operations? (Is new functionality required?).

The third and fourth questions are central to the argument. In this chapter we do not take the view that SDA is simply a collection of techniques – a sequence of 'mechanical' operations, if you will. The view here is that it is important to distinguish between the conduct of analysis and the tools (or techniques) employed in data analysis. SDA should be seen as an 'open ended' sequence of operations with the aim of extracting information from data. To do this, SDA employs a variety of tools (some of which are specialist because of the nature of spatial data (Haining, 1990, pp. 40–50; Anselin, 1990)). These tools are 'closed' in the sense that they are based on the application of formulae interpreted through usually well-defined decision criteria. It is the tools that can be implemented in the form of computer algorithms. Hence the claim here is that SDA should be defined as the subjective, imaginative ('open ended') employment of analytical ('closed') tools on statistical data relating to geographical events in order to achieve one or more of the aims of description, understanding, prediction or control (Wise and Haining, 1991). This issue is felt to be of importance when considering the design of SDA modules and what tools need to be available to enable the analyst to carry out a coherent programme of SDA.

The incorporation of SDA into GIS will extend GIS utility and will provide benefits to the academic community where the current GIS functionality (focusing on primitive geometric operations, simple query and summary descriptions) is seen as very limited.[3] There are, however, other interested parties for whom GIS would be of more interest were it to have this extended capability. In the United Kingdom, Area Health Authorities are showing an interest in GIS for handling and analysing extensive health records for purposes of detecting and explaining geographical patterns of health events. The points raised in this chapter will be illustrated using analyses of intra-urban cancer mortality data for a single time period.

In the next section we consider the first two questions and consider the implications for the analysis of health data. Section 3 examines the conduct of SDA while section 4 discusses the set of SDA tools required to support one type of coherent programme of data analytic research on health data. In section 5 these tools are abstracted into a set of fundamental operations.

What types of data? What types of questions?

Geographic Information Systems

For information about the world to be stored in digital databases, reality must be abstracted and in particular, discretized into a finite, and small enough, number of logical data units. This process is termed data modelling. From the GIS perspective reality is conceptualized in one of two ways (Goodchild, 1992). *Either* the world is depicted as a set of layers (or fields) each defining the continuous or quasi continuous spatial variation of one variable, *or* the world is depicted as an empty space populated by objects. The process of discretizing a reality viewed as a set of fields leads to attribute values being attached to regular or irregular distributions of sample points, a collection of contour lines (joining points of equal attribute value) or a regional partition (consisting of areas and lines). In the case of a regional partition, different assumptions can be made about the form of intraregional spatial variation in attribute values. Boundaries may be 'natural', coinciding with real changes in the field variable (such as break of slope, lines of discontinuity), or be 'artificial' in the sense of being imposed independently of properties of the field variable (such as administrative boundaries). The object view of reality is based on the representation of reality as collections of points, lines and areas with attribute values attached. It is evident that through the process of discretization, quite different conceptualizations of reality might be given identical representations in the database. Points for example may refer either to a field or an object based conceptualization of reality. Also, identical real world features might be given different object representations in different databases, e.g. a town represented either as a point or as an area. This is often because of one or more of: the quality of the source data, the scale of the representation, the purpose for which the database was constructed.

An object view of reality generates two primary classes of question: questions relating to the objects, and questions relating to the properties of independently defined attributes that are attached to the objects. Objects may be embedded in what is conceptualized as a one-dimensioned space (such as a road or river) or they may be embedded in a two-dimensional (regional) space. A univariate analysis treats one type of object class at a time (points or lines in a linear space; points, lines or areas in a two dimensional space). A multivariate analysis treats two or more types of objects simultaneously that may be either of the same class, e.g. point – point, or not, e.g. point-line, with the purpose of identifying relationships between the sets of objects. An example of the former is an analysis of the distribution of settlements in relation to point sources of water, an example of the latter is an analysis of the distribution of settlements in relation to river networks.

Questions that focus on the objects themselves may be further classified in terms of whether they refer to locational properties, e.g. where the objects are on the map and the spatial relationships between them, or whether they refer to geometrical issues, e.g. point density; line length and direction; area size and shape. While the former are intrinsically spatial, questions concerning geometrical properties may be aspatial, e.g. 'what is the frequency distribution of area size?', or spatial, e.g. 'what is the variation in the distribution of area size across the region or between different parts of the regional space?'

Questions that focus on independently defined attributes that are attached to the objects can be classified into whether they relate to a single attribute or more than one attribute (univariate or multivariate attribute analysis) and whether they relate to a single object class or more than one object class (univariate or multivariate object class analysis). These questions can be further classified into whether they are aspatial, e.g. 'is there a significant correlation between attributes A and B across the set of areas?', or spatial, e.g. are large/small values of attribute A spatially close to large/small values of attribute B across the set of areas?' It appears to be the case that for any aspatial query there is a spatial query in the sense that we can ask how data properties vary across the map or between one part of the map and another.

In the case of the field representation of reality there appears to be two primary classes of questions. In the case of point sampled surfaces one class of question is to interpolate values either to other, specified, point sites (where no observation was recorded) or to the whole region for purposes of mapping. The second primary class of question concerns the properties of the attributes attached to the points, lines or areas of the discretization. These may be classified in terms of whether a univariate or multivariate analysis is required, whether the questions are aspatial, e.g. 'what is the mean value of the attribute over the region?', or spatial, e.g. 'are there spatial trends in the value of the attribute over the region?'

The objects generated by the discretization of a field are of interest but in a different sense to that encountered in the case of object based representations of reality. Discretizing a surface involves a loss of information and moreover the objects of the discretization are not observable features in the world in the same sense as in an object based representation of reality. So, in order to answer the primary questions it is usually necessary to try to assess the sensitivity of findings to the form of the (sometimes arbitrary) discretization. Interpolation errors are a function of the density and spacing of point samples; relationships between attributes are a function of the size of the regional partition and the relationship of that partition to the underlying surface variability.

Figures 3.1 and 3.2 summarize the classes of questions arising from object and field based views of reality.

In conclusion, note that this section has dealt with the questions asked of object and field based data sets rather than the forms of analysis and has distinguished between aspatial and spatial *queries*. It is important to bear in mind, in the case of spatially referenced data, that in order to provide a rigorous answer to an aspatial as well as a spatial query it may be necessary

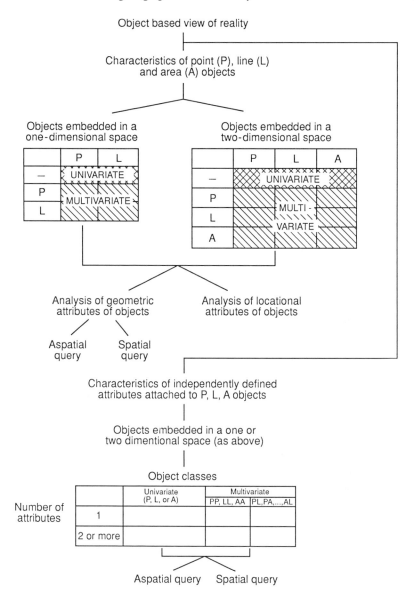

Figure 3.1. Classes of questions arising from an object view of reality.

R. Haining

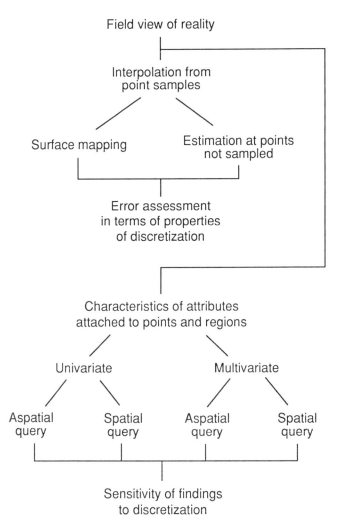

Figure 3.2. Classes of questions arising from a field view of reality.

to use the specialist tools and methods of spatial data analysis rather than relying on the tools of standard statistics. We shall expand on this point in a later section.

Intra-urban cancer mortality

There are two important data sources that underpin UK geographical research on intra-urban variation in cancer mortality: the census and the cancer registry. The cancer registry lists, among other things, the home addresses of cancer patients. Complete digitized listings of all addresses in a city can be purchased (patients and non-patients). Were such a database to be constructed it would reflect an object based representation of cancer incidence, with points representing individual addresses with attributes specifying whether individuals at that address were on the registry or not, and if so the diagnosis, date of death (if it has occurred) and some personal details (e.g. occupation) etc. For several reasons however (not least the size of such a database) a field-based representation is usually adopted. First, for reasons of confidentiality, patients are usually identified only by a five to seven character (unit) postcode rather than the full address. Second, if the aim is to merge incidence data with relevant information on socioeconomic variables from the census, these data are only available in terms of enumeration districts (EDs). EDs are larger than the unit postcodes. In a city the size of Sheffield (population: 500,000) there are about 11,000 unit postcodes (of which 8,900 are residential) and about 1,100 EDs.[4] Although ED boundaries are clearly defined (so they can be digitized), the criteria employed are qualitative and wide ranging (Evans, 1981) with no guarantee either of intra-ED homogeneity (in terms of socio-economic variables) or spatial compactness.[5]

If the purpose of analysis is a descriptive analysis or test of clustering on the incidence data alone a database constructed from a field based representation using unit postcodes as the primary spatial units is possible but assumptions have to be made about population levels in each unit postcode (Alexander *et al.*, 1991). If the purpose of analysis is to relate incidence data to population characteristics including socio-economic characteristics EDs become the primary spatial units. However there are complications. The cancer data need to be matched to EDs and converted to standardized rates (such as the Standardized Mortality Ratio, adjusted for the size, age and gender composition of each ED). Socio-economic variables will also need standardizing (converted to percentages of the ED population or numbers of households). But for areas as small as EDs, many will have no cancer cases (while the rest will have only one or two cases) and all rates and percentages will be sensitive to individual occurencies and recording errors. For these reasons a further level of aggregation and data standardization is desirable. EDs need to be merged without greatly diminishing such intra-area homogeneity as the individual EDs possess. Variables that show strong spatial persistence (autocorrelation) will not suffer from a significant loss of information under such an operation if aggregation does not go beyond the scale at which strong autocorrelation is present. Classification routines have been developed to construct regional aggregation of EDs where several variables are used as the basis for the aggregation (see Semple and Green, 1984).

Figure 3.3 (a) identifies the data models and Figure 3.3 (b) identifies a set of interrelated questions or queries that arise in geographical health data research focusing on the description and modelling of cancer rates (mortality, incidence, prevalence).

Section 4 of this chapter describes some of the tools that may be useful in tackling these questions. The next section, however, outlines some important

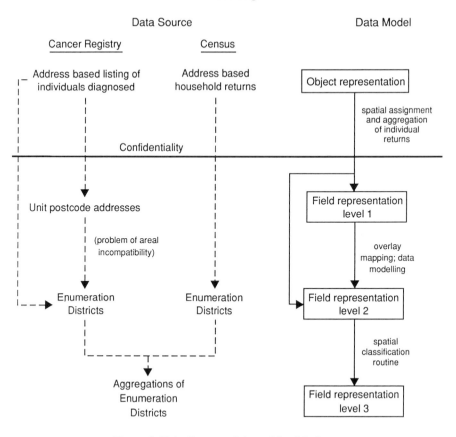

Figure 3.3(a). Data models and health data.

approaches to the way in which SDA is conducted. These provide contexts within which individual tools (and sets of tools) are used.

The conduct of SDA

This section describes the conduct of SDA from two points of view: one a user/data oriented perspective, the other a statistical perspective. We start with the user/data oriented perspective and make use of a simple classification which is summarized in Figure 3.4. The distinction on the user axis is between the 'expert' data analyst whose expertise lies in handling data and extracting information from data, with a sound understanding of statistical method, and the 'non expert' data analyst who has a need to extract information from data but who would not claim a strong grounding in statistical method at least as it relates to statistical data modelling and inference. The distinction on the data axis is between low quality data which is of unknown reliability but often available in large amounts, and high quality data which

Field	Query			
Level	Univariate		Multivariate	
	Aspatial	Spatial	Aspatial	Spatial
Unit Postcodes	1) Describe the properties of the set of disease incidence values a) Numerical b) Graphical	1) Map cases 2) Is there evidence of clustering ?		1) Is there evidence of clustering in respect of specified sources or possible causes ?
EDs	1) As above	1) As above 2) As above		1) As above
Aggregated EDs	1) As above 2) Is there evidence of high concentrations of cases ?	1) Map rates 2) Describe spatial distributional properties 3) Evidence of trends, patterns, other spatial variations 4) Identify extreme areas	1) What relationships exist between incidence rates and socio-economic variables ? 2) Can we model these relationships ? 3) What areas have extreme high or low rates ?	1) What spatial associations exist between incidence rates and other variables ? 2) Would a spatial model describe relationships better ?
		Sensitivity of findings to areal partition, map boundaries etc.		Sensitivity of findings to areal partition, map boundaries etc.

Figure 3.3(b). Data analysis questions in health research.

may or may not be available in large amounts but is considered reliable in part because of the time, effort and expense incurred in acquiring and checking it.[6] Both types of data are important and we assume the best available for what the analyst has in mind.

Within the context of this classification we can make general, preliminary, observations on the aims of data analysis and what is required if these aims are to be fulfilled. The expert user is likely to want some type of inference framework in order to evaluate his or her models whereas the non-expert user will largely be concerned with summarizing properties and using simple descriptive techniques (e.g. scatterplots, correlation coefficients) to explore possible relationships. While the non-expert user might be satisfied with a

USER

		Non-Expert	Expert
DATA QUALITY	Low	Aim : Detect patterns, simple relationships. Summarize data properties. Need : Simple robust descriptive statistical tools	Aim : As for non-expert but may also want to explore data errors, identify outliers, assess sensitivity Need : Simple robust descriptive statistical tools. Robust model fitting tools. Combinatorial inference framework
	High	Aim : As above Need : As above	Aim : As above, but may want to engage in more extensive data modelling and hypothesis testing. Need : As above, but also confirmatory tools including an inference framework for sifting variables, choosing between models and assessing model fit.

Figure 3.4. A user/data perspective on the conduct of SDA.

Exploratory / Descriptive   Confirmatory / Inferential

Aim : Identify data properties   Model building; evidence assessment

Needs : Robust methods to summarize   Methods to aid model specification,
          data for purposes of identifying   parameter estimation, hypothesis
          data properties and pattern   testing and model validation
          detection.   i) Data driven approaches
              Numerical methods   ii) Model driven approaches
              Graphical methods
              Cartographical methods

Figure 3.5. A statistical perspective on the conduct of SDA.

blanket qualifier on findings when working with low quality data, the expert is more likely to want to explore the sensitivity of findings to known short-comings in the data. Where large volumes of data are available there may be problems in implementing some standard tools and fitting certain models to spatial data. This is an important issue but not one that can be covered here (see, for example, Cliff and Ord, 1981; Haining, 1990).

A statistical perspective on the conduct of SDA is given in Figure 3.5. The contrast between exploratory and confirmatory approaches is not exclusive. Exploratory methods may be of great value in identifying data properties for purposes of model specification and subsequent model validation in a pro-gramme of confirmatory analysis. Nonetheless this appears to be a useful distinction to draw not least because it draws a line between those areas of applied statistical analysis which are concerned with data description and 'letting the data speak for themselves' from those areas of applied statistical analysis where the goal is inference, data modelling and the use of statistical method for confirming or refuting substantive theory. Given the user/data perspective described above this division also reflects the non-expert/expert user division which some feel is an important one within the context of GIS (e.g. Openshaw, 1990). Given the lack of underlying substantive theory in many of the areas of current GIS application it can be argued that some of the paraphernalia of confirmatory analysis, particularly that which has been developed as part of the 'model driven' philosophy of data analysis is not relevant here.[7] The argument in this chapter, however, is that GIS offers possibilities for a range of users. If a collection of exploratory/descriptive tools would be of benefit to virtually all users, there needs also to be made available some confirmatory/inferential tools in order to provide an adequate toolbox for a coherent programme of 'expert' data analysis that goes beyond description. In the next section we specify the content of such a statistical toolbox, again in the context of intra-urban health data research.

Requirements for SDA

Haining (1990, 1991a) reports the results of an SDA of cancer mortality data for the city of Glasgow (1980–2), based on 87 community medicine areas (CMAs), which are aggregations of EDs. Data analysis was concerned with three linked activities: (1) to describe variations in cancer mortality across the 87 CMAs; (2) to model variations in cancer mortality rates in terms of socio-economic variables; (3) to identify CMAs with particularly high or low levels of mortality. The question of the sensitivity of findings to the areal system was raised but not addressed.

The mortality data values were converted to standardized mortality rates (SMRs). The usual standard numerical and graphical statistical summary measures were computed: measures of central tendency (mean, median, mode) measures of dispersion (variance, standard deviation, quartiles) and graphical representations of the distribution of values (histograms, stem and leaf, box plots). Both 'conventional' and resistant summary statistics are useful. The quartile values can be used to identify SMRs that are extreme with respect to the distribution of values. The criterion employed is that a value (y) for the ith CMA is extreme (called an outlier) if either

$$y_i > F_U + 1.5 \ (F_U - F_L) \ \text{or} \ y_i < F_L - 1.5 \ (F_U - F_L)$$

where F_U and F_L are the upper and lower quartiles and $(F_U - F_L)$ is called the interquartile range. In all these cases, nothing needs to be known about the location of values on the map of Glasgow and in the context of Figure 3.2 these analyses relate to aspatial univariate data queries. In addition to these, maps were drawn of the SMRs and scatterplots of SMR against distance from the centre were obtained. Both the plots are designed to detect trends in rates. In addition a plot of rates for CMA i (y_i) against the average of the rates in the adjacent CMAs ($(Wy)_i$) were also plotted in order to explore local scales of pattern. A regression of y_i on $(Wy)_i$ was computed to detect 'spatial' outliers. In the context of the reported research these plots provided useful preliminary spatial information as well as material that was helpful at later modelling stages of the analysis. These analyses, following Figure 3.2, relate to the category of spatial univariate data queries.

The problem for the analyst is that none of the spatial univariate data queries can be easily answered using a standard statistical package (like SPSS or MINITAB) and any individual cases highlighted by the analyst cannot be readily transported to a map display. Such a facility (e.g. for identified outliers) would be of considerable value. Nor is it easy to examine spatial subsets of the data for purposes of detecting heterogeneity. However there is more than just a technical issue here in making appropriate software available for a fully spatialized exploratory/descriptive data analysis. As noted in Cressie (1984), Haining (1990) and Anselin and Getis (1992) many of the current methods for exploratory data analysis (EDA) assume independence of observations so their validity when applied to spatial data is still open to question. As Anselin and Getis (1992, p. 6) remark: 'a true "spatial" EDA . . . does not exist'. This remark applies not only to spatial queries (e.g. what are the outliers on a scatterplot of y_i against $(Wy)_i$) but also to aspatial queries. For example, the D statistic can be used to test for the presence of outliers in a set of n observations. Now

$$D = n^{1/2}(\bar{y} - \text{median } (y))/(\sigma(0.5708))^{1/2}$$

where $\bar{y}$ is the sample mean and $\sigma = (F_U - F_L)/1.349$. If assessed as a standard normal deviate this statistic assumes independent observations.

Multivariate analysis addresses questions concerning the relationship between mortality rates and socio-economic relationships. A simple aspatial query is to check for association between y_i and each socio-economic variable through scatterplots and then, more formally through bivariate correlation. Ostensibly correlation between two variables could be examined by using a standard package. In the case of spatial data, although the Pearson, or Spearman, correlation coefficients are computed in the normal way, any test of significance has to allow for the information loss spatial dependence introduces and for variation in the size of the areal units (heterogeneity of units). The theory is summarized, together with a worked example, in Haining (1991b).

Regression modelling of mortality data uses the model:

$$Z_i = \beta_o + \sum_j \beta_j X_{ji} + \eta_i$$

where Z_i is the log transformation of y_i and $\{X_{ji}\}_j$ are the set of explanatory socio-economic variables. The errors $\{\eta_i\}$ are, under the assumption of independent Poisson counts for the number of deaths in region i, normally distributed with a zero mean and variance $\sigma^2\{b_{ii}\}$ where $b_{ii} = [1 + (\sigma^2 0_i)]^{-1}$. 0_i is the observed number of deaths in the ith CMA. Derivations and details of the fitting procedure are given in Pocock *et al.* (1981) and reviewed in Haining (1991a).

The multiple regression model is, like the correlation coefficient, an aspatial query in the sense that to fit the model nothing need be known about the spatial referencing of the data. However, there are a number of aspects of the fitting of this regression model that, again, cannot be handled readily in a standard statistical package:

1. Residuals $\{\hat{\eta}_i\}$ must be checked for independence. Given that the areas are adjacent there is the possibility of spatial autocorrelation in the residuals. This would occur if, for example, an important explanatory variable were excluded from the regression model which was spatially autocorrelated. Scatterplots of η_i against $(W\hat{\eta})_i$ and more formal tests for spatial auto-correlation depend on having information on the spatial location of each case. (See, for example, Anselin (1988), who shows that problems of spatial autocorrelation and heterogeneity induced by different area sizes require specialist tests.)
2. If spatial autocorrelation is found in the model residuals, the analyst might wish to respecify by incorporating trend surface or dummy variables to allow for spatial differences within the area.
3. In the case of an infectious disease levels of mortality in one area might be influenced by levels of mortality in adjacent areas and the analyst might want to fit a regression model with a spatial lagged form of the dependent variable as an explanatory variable.[8]
4. To reflect population mobility across the CMAs the analyst might wish to respecify the regression model introducing spatially lagged forms of some of the independent variables as additional explanatory variables.[9]
5. In some or all of the above models the analyst might need to consider the effects of study area boundaries as well as the spatial distribution of observations and the spatial distribution of outliers in assessing the model fit. The analyst might wish to delete cases and refit the model.

This selection of problems reveals a central issue. The special nature of spatial data affects model specification and model validation. Furthermore, since it affects model specification it affects model estimation (Ord, 1975). What might start as an aspatial query cannot be answered rigorously using a standard statistical software package because of the special properties of spatial data in relation to the assumptions underlying standard statistical theory and usually implicit in such packages. It may be necessary to carry out special checks that are not readily handled in such packages and to develop more explicitly spatial models for the data.

Figure 3.6 summarizes the main statistical tools employed in the analysis of the Glasgow mortality data.

Tools for data manipulation: Standardization (e.g. percentages, SMR's)
Transformations (e.g. logarithmic)

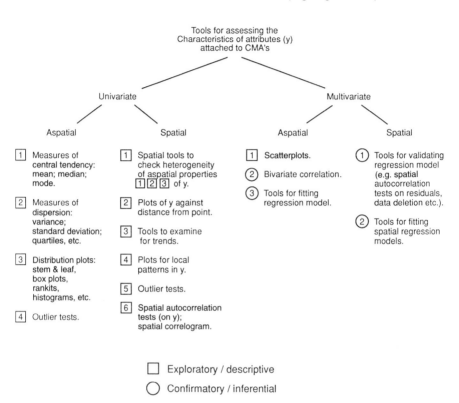

Figure 3.6. Tools for SDA of urban mortality data.

Fundamental spatial operations for SDA tools

In this section, fundamental spatial operations are specified that appear to underlie the specialist SDA tools that were referenced in the previous section. The implication is that within GIS it ought to be possible to carry out all these operations relatively easily. These operations are classified here, as follows:

[A] Operations that depend on accessing the location and/or attribute values of cases.
[B] Operations that depend on identifying spatial relationships between cases.
[C] Operations that involve interaction between type [A] and [B] operations.

A summary is given in Figure 3.7.

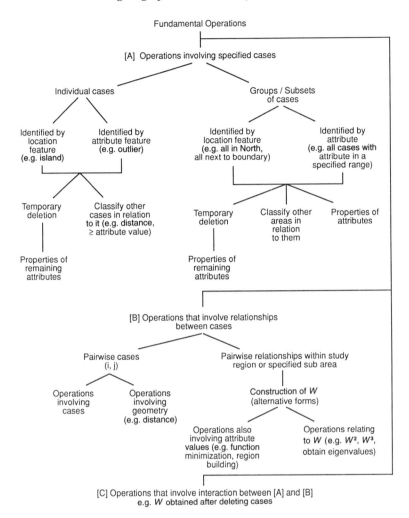

Figure 3.7. A classification of fundamental operations for SDA tools.

Type [A] operations

These operations are distinguished by the fact that the analyst can identify individual data cases or sets of cases either by systematically highlighting which cases are to be treated separately or by defining location or attribute criteria. In the situation where only a single case is to be treated separately it might be an 'island' outlier, separate from the rest of the study region or it might possess an extreme attribute value. In either case the analyst may wish to recompute on the basis of the reduced data set in order to assess the sensitivity of results to individual cases. In the situation where area subsets are highlighted for special treatment then the purpose might be to examine the data for spatial heterogeneity.

The distance plots and 'lag' distance box plots in the Glasgow study would be implemented by the analyst identifying the central zone and computing each area's distance away or classifying each area by its 'lag order' relationship to the central zone. Other aspects of the analysis involved contrasts between areas to the North and South of the River Clyde, between suburban and inner city areas. Such analyses would be made possible if operations like those described above were readily available within a GIS. One important development in this area is the work of Haslett *et al.* (1990). Using interactive linked windows they have developed software that links descriptive summaries of spatial data to the location of cases on the corresponding map.

Type [B] operations

If type [A] operations relate principally to locational characteristics of areas, type [B] operations relate principally to relational characteristics between cases. The facility to explore relational attributes is fundamental to a spatial analysis whether the data queries are spatial or aspatial.

A distinction is drawn in Figure 3.7 between operations that are built up from pairwise comparisons and those where the full set of relations are needed as captured in the connectivity or weights matrix (*W*) introduced by Cliff and Ord (1981) for handling order relationships between data points (or areas) in space.

For many spatial data analysis techniques, pairwise relationships between points may be all that is required in order to perform the necessary computations. This seems to be true for example of all the spatial autocorrelation statistics. With these statistics it is sufficient to identify neighbours and then compute attribute differences squared (the Geary coefficient) or attribute products (the Moran coefficient). It may be easier to compute these coefficients after first constructing and storing the full *W* matrix but it does not seem to be necessary.

Other forms of SDA require the construction and storage of *W* since analysis involves operating on the full spatial system. This is probably necessary for region building classification procedures (for exploring the sensitivity of findings to alternative spatial partitions). It is certainly necessary to work with *W* for computing eigenvalues, matrix powers and other matrix properties used in the process of fitting spatial models (Haining, 1990). Ding and Fotheringham (1992) have developed a module in ARC/INFO to construct *W* matrices exploiting the left-right topology associated with the lines defining area boundaries in a vector database. It will be useful also to have the facility to compute *W* matrices based on other properties of the spatial system such as length of shared common boundary as well as independent flow or interaction data (see Haining, 1990, pp. 69–74).

Type [C] operations

The third category includes operations that simultaneously draw together type [A] and [B] operations. If it is to be possible to assess the sensitivity of modelling results to data deletion then the analyst will want to be able to select cases for deletion and be able to restructure automatically the *W* matrix (Martin, 1984).

The classification suggested here is specific to SDA and 'application' driven. Nonetheless these activities appear to be basic to current SDA. Goodchild (1987) in a more general context, has identified six basic classes of SA operations. As defined here, type [A] operations appear to fall within Goodchild's class (1) and (2) operations while type [B] and [C] fall within his class (3) and (4) operations. The above discussion perhaps goes some way to opening up Goodchild's classes of operations in order to identify how they contribute to performing SDA.

Conclusion

To make SDA techniques available in the form of well-tested software is essential if these techniques are to benefit the wider scientific community. GIS offers a potentially valuable platform for SDA software helping to place SDA techniques where they belong, within the grasp of those for whom the methodology is being developed, namely data analysts with substantive problems to address.

We have argued here that at this early stage it is important to think about the tools that are necessary to go in software packages so that the analyst is in a position to undertake a coherent programme of work. Piecemeal technical increments will undoubtedly carry the research effort forward but it is also important to consider the wider functionality that is needed to support that larger package of tools. What is the set of tools (in general terms or in relation to specific classes of problems) and to what extent is the necessary functionality already available, or could be made available? To what extent is it necessary to engage in a reappraisal of the structure of GIS databases?

The discussion here has been in terms of intraurban mortality data analysis but the methodology of obtaining descriptive summaries and proceeding to multivariate regression modelling is of such fundamental importance that what has been discussed here is relevant to many other applications.

Acknowledgement

The author wishes to thank Stephen Wise for many helpful discussions on issues linked to this chapter.

Notes

1. Goodchild (in Haining and Wise, 1991) argues that the benefits to SDA of closer linkage with GIS extends beyond the practicalities of doing SDA. Working in a GIS environment will force analysts to think more about the nature of the underlying data model and hence what questions it is sensible to ask of any given data set or class of data. In fact the definition of an 'event' for purposes of SDA matches well with the data models through which geographic reality is represented in digital databases (see section 2).
2. The form of the linkage between any GIS and SDA software is not addressed here. These issues are discussed for example in the contributions by Goodchild, Wise, Bossard and Green in Haining and Wise (1991).
3. The Sheffield ESRC sponsored workshop (Haining and Wise, 1991) focused also on aspects of SDA that would enhance *current* GIS practice including reconciling data sets from different spatial partitions, estimating missing values and spatial sampling.
4. See Evans (1981) for details of UK census data collection. Unit postcodes and EDs do not match which may create a problem of incompatible areal units when attempting to merge the two data sets (Flowerdew and Green, 1989).
5. Grid references for unit postcodes are available in the form of digitized point locations (e.g. see Wise, 1991).
6. Spatial data may be unreliable in two senses. There may be errors in the attribute values and errors in the locational assignment of attribute values (see, for example, Chrisman, 1987).
7. For a discussion of 'data driven' and 'model driven' approaches to data analysis see, for example, Anselin (1988), Anselin and Getis (1991) and Haining (1990). A paper by Gilbert (1986) provides an excellent introduction to an important distinction within the 'model driven' econometric methodology. In the geographical literature the term 'data driven' is sometimes equated with 'spatial statistics' and the term 'model driven' with 'spatial econometrics' since it is the economic/regional science branch of geography where models have been developed that can benefit from such a rigorous methodology.
8. The term 'lag' is used in the econometric sense. First spatial lag neighbours of a CMA are those CMAs that are adjacent. The term is discussed in more detail in Haining (1990, pp. 69–74).
9. Unlike laboratory experiments consisting of n discrete replications, a spatial system consists of n sets of observations but no assurance that outcomes in any one region are solely a function of conditions in the same region.

References

Alexander, F. E., Ricketts, T. J., Williams, J. and Cartwright, R. A., 1991, Methods of mapping and identifying small clusters of rare diseases with applications to geographical epidemiology, *Geographical Analysis*, **23**, 158–173.

Anselin, L. and Getis, A., 1991, Spatial statistical analysis and Geographic Information Systems, Paper presented at the 31st European Congress of the Regional Science Association, Lisbon, Portugal.

Anselin, L., 1988, Lagrange multiplier test diagnostics for spatial dependence and spatial heterogeneity, *Geographical Analysis*, **20**, 1–17.

Anselin, L., 1988, *Spatial Econometrics, Methods and Models*, Dordrecht: Kluwer Academic Press.

Anselin, L., 1990, What is special about spatial data? D. Griffith (Ed.), *Spatial Statistics: Past, Present and Future*, Institute of Mathematical Geography, Michigan Document Services, Michigan, pp. 61–77.

Chorley, R. J., 1972, *Spatial Analysis in Geomorphology*, London: Methuen.

Chrisman, N. R., 1987, The accuracy of map analysis: a reassessment *Landscape and Urban Planning*, **14**, 427–439.

Cliff, A. D. and Ord, I. K., 1981, *Spatial Processes: Models and Applications*, London: Pion.

Cressie, N., 1984, Towards resistant geostatistics, G. Verly, M. David, A. G. Journel, A. Marechal (Eds.), *Geostatistics for Natural Resource Characterisation*. Dordrecht: Reidel, pp. 21–44.

Ding and Fotheringham, A. S., 1992, The integration of spatial analysis and GIS, *Computers, Environment and Urban Systems*, **16**, 3–19.

Evans, I. S., 1981, Census Data Handling, N. Wrigley and R. J. Bennett (Eds.), *Quantitative Geography*, Routledge and Kegan Paul, pp. 46–59.

Flowerdrew, R. and Green, M., 1989, Statistical methods for inference between incompatible zonal systems, M. F. Goodchild and S. Gopal *Accuracy of Spatial Databases*, London: Taylor and Francis, pp. 239–247.

Gilbert, C. L., 1986, Professor Hendry's econometric methodology, *Oxford Bulletin of Economics and Statistics*, **48**, 283–307.

Goodchild, M. F., 1987, A spatial analytical perspective on geographical information systems, *International Journal of Geographical Information Systems*, **1**, 327–334.

Goodchild, M. F., 1992, Geographical data modelling, *Computers and Geosciences*, **18**, 401–8.

Haining, R. P., 1990, *Spatial Data Analysis in the Social and Environmental Science*, Cambridge: Cambridge University Press.

Haining, R. P., 1991a, Estimation with heteroscedastic and correlated errors: a spatial analysis of intra-urban mortality data, *Papers in Regional Science*, **70**, 223–41.

Haining, R. P., 1991b, Bivariate correlation with spatial data, *Geographical Analysis*, **23**, 210–227.

Haining, R. P. and Wise, S. M., 1991, *GIS and Spatial Data Analysis: Report on the Sheffield Workshop*. Regional Research Laboratory Discussion Paper, 11 Department of Town and Regional Planning, University of Sheffield.

Haslett, J., Bradley, R., Graig, P. S., Wills, G. and Unwin, A. R., 1991, Dynamic Graphics for exploring spatial data with application to locating global and local anomalies, *American Statistician*, **45**, 234–242.

Martin, R. J., 1984, Exact maximum likelihood for incomplete data from a correlated Gaussian process, *Communications in Statistics: Theory and Methods*, **13**, 1275–88.

Openshaw, S., 1990, *A Spatial Analysis Research Strategy for the Regional Research Laboratory Initiative*, Regional Research Laboratory Discussion Paper 3, Department of Town and Regional Planning, University of Sheffield.

Ord, J. K., 1975, Estimation methods for models of spatial interaction, *Journal, American Statistical Association*, **70**, 120–126.

Pocock, S. J., Cook, D. G. and Beresford, S. A. A., 1981, Regression of area mortality rates on explanatory variables: what weighting is appropriate? *Applied Statistician*, **30**, 286–296.

Semple, R. K. and Green, M. B., 1984, Classification in Human Geography, G. L. Gaile and C. J. Willmott (Eds.), *Spatial Statistics and Spatial Models*, Dordrecht: Reidel, pp. 55–79.

Wise, S. M., 1991, Setting up an on-line service to access the Postzon file, *Wales and S. W. Regional Research Laboratory Technical Report* 32.

Wise, S. M. and Haining, R. P., 1991, The role of spatial analysis in geographical information systems, *Proceedings AGI 91*, London: Westrade Fairs, 3.24.1–3.24.8.

4

Spatial analysis and GIS

Morton E. O'Kelly

Introduction

Many of us involved in spatial analysis research are excited to see the explosive growth of interest in GIS. It is clear that the rapid growth of GIS has given a big boost to fundamental research in spatial analysis, and in many ways *solidifies* the future of the quantitative focus of the discipline.

There exist potential linkages between many aspects of spatial analysis and the new information processing, data handling, data storage and data display techniques available through GIS. Furthermore, there are strong emerging links to applications in other disciplines with parallel interests in spatial analysis (including ecology, archaeology, natural resources, landscape architecture and geodetic science, etc.) These allied fields have long used spatial analysis techniques (see, for example, Upham, 1979; Bartlett, 1975; and Diggle, 1983) but are now increasingly using GIS as a creative tool.

Against the background of this growth, and from the perspective of a spatial analyst looking out towards the next stages of development in GIS, two major directions which need attention are apparent. First, the traditional methods for displaying data about spatial situations, and for presenting the results of spatial analyses, need to be overhauled in view of the great advances in information processing technology. This line of attack is straightforward: it suggests that new display techniques be added to the output of existing spatial analysis operations. Ideally, the analyst would use the GIS in query mode (Goodchild, 1987) to develop an improved understanding of the properties of a spatial system. While this is easy to state in principle, the actual implementation will stretch current capabilities and will require a rethinking of the conceptual bases for spatial analysis. One such example might arise if improved display techniques increased the accessibility of multiobjective programming, thereby encouraging analysts to take a multi-objective view, and therefore replacing existing unidimensional methods. One obvious area for this to play a role is in locational analysis.

Second, spatial analysts need to help GIS end-users and providers to understand and to improve their sets of tools, and to enhance the appropriate levels of theory and modelling capability in real problem-solving situations. This need is particularly acute if *appropriate* analytical methods are to lie at the foundation of developments in GIS. The capability to adapt existing algorithms to the data structures in GIS is a critical component of this research.

The overall theme of this chapter is that when GIS integrates spatial analysis at a fundamental level, the full potential of spatial analysis will be
unlocked. The case is made by demonstrating that there are potential *new*
results and advances that are obtainable by linking spatial analysis and GIS.
In the style of a position paper, this effort builds the case for an improved
linkage between spatial analysis and GIS by showing that GIS can be improved
by adding innovative spatial analysis functions, and in turn by showing how
certain spatial analysis operations are enhanced using GIS. The third section
of the chapter mentions some of the immediate incremental building blocks
that are needed to start this transformation of spatial analytical methods,
while the concluding section mentions several barriers to the realization of
these aims.

Innovations in spatial statistics and GIS applications

As discussed in the previous section, the next few years promise an unpre-
cedented opportunity for spatial analysts to promote and contribute to the
spread of new tools to the GIS community. This section is designed to make
some of the power of these types of tools apparent to the potential user. To
attempt to make the discussion more concrete, let us go back to some fundamental building blocks or entities in spatial analysis: i.e. points, lines and
areas. Associated with each of these classes of entities are new spatial analytical operations that need to be developed. Couclelis (1991) correctly points
out that the focus on space as a container does not provide the platform for
answering the complex types of spatial questions posed in planning; however,
see more on the site and situation distinction she draws in later paragraphs.
In each case the role of new exploratory tools, visualization, and space-time
analyses can be seen in slightly different ways.

Space-time pattern recognition in point data sets

Geographers have been slow to integrate *both* temporal and spatial dimension into GIS. There are formidable technical obstacles, but even at the
fundamental research level, analysts have been unable to operationalize the
role of space and time in simple models. A concrete example may help. A
conceptual model of space and time acting as a constraint on human activity
(Hagerstrand) has gone largely untested because of the difficulty of translat-
ing activity records from travel diaries into three dimensional diagrams called
'prisms' by Hagerstrand (however, see Miller, 1991). There have been studies
of the importance of distance (Gatrell, 1983) and time (Parkes and Thrift,
1980; Goodchild and Janelle, 1984; and Janelle *et al.*, 1988), but despite these
impressive research efforts, it is uncommon to consider combined time-space

'reach' as a quantitative tool in measuring transportation plans. Two exceptions include Villoria's (1989) dissertation, which measured the size of activity fields in an urban case study in the Philippines, and Kent's (1984) mapping of activity spaces from archaeological data.

A model for space-time data analysis

Assume that the events are located at distinguishable locations, and that a simple date is used to keep track of the time of events. Let $W_{ij} = f(t_i, t_j)$ where t_i and t_j are the times of the events i and j, and further assume that i and j take place at $p_i = (x_i, y_i)$ and $p_j = (x_j, y_j)$ respectively. Supposing that W_{ij} is a decreasing function of the time between the events

$$W_{ij} = 1 / \{e + | t_i - t_j |\}.$$

Then, given thresholds D and T, classification of events in space and time can be simply represented in a two-by-two table (Table 4.1).

	SPACE	
TIME	$d(p_i, p_j) < D$	$d(p_i, p_j) > D$
$W_{ij} > T$ highly interactive	CLOSE TOGETHER CONTEMPORANEOUS	FAR APART CONTEMPORANEOUS
$W_{ij} < T$ weakly interactive	CLOSE TOGETHER TIME GAP	FAR APART TIME GAP

The statistical analysis of such 2-way tables is a fundamental one for recognizing whether or not there are significant patterns in space and time (see Knox, 1964). The strength of the temporal interaction is measured as follows: W increases as the events are closer together in time, and D increases as the events are further apart in space. To what extent are observations which are close together in time also spatially clustered? The problem is to recognize clusters or groups in the set. While there are many follow-up papers to Knox's pioneering analysis, the method is problematic because of the non-independence of events (see Glick, 1979). The next section develops a novel mathematical approach to the problem of space-time pattern analysis: one which is of great potential usefulness in identifying the spatial pattern and perhaps helping to uncover the underlying spatial process.

Consider a set of n interacting points in a two-dimensional space. The levels of interactions between the observations are given exogenously, as functions of their temporal separation. Assume that the cluster means must be adjusted to reflect the interaction between the entities. For example, consider a system of $n = n_1 + n_2$ nodes such that the n_1 subset is temporally linked, and the n_2 subset is also highly interactive among themselves, and for the sake of illustration suppose that there are negligible interactions between the subsets. (That is the $(n_1 \times n_2)$ and $(n_2 \times n_1)$ subsystems contain only zero interactions.) If the n_1 and n_2 nodes are plotted graphically, it would be fortuitous if all the n_1 and n_2 nodes could be separated neatly into two easily identifiable spatial groups. Indeed, since there is no requirement of contiguity for the interacting entities, there is no guarantee that a cluster of points n_1 should contain only adjacent nodes. While the conventional geostatistical clustering problem for several groups yields a partition with the property that all the observations which are closer to centroid A than to centroid B are assigned to the same group, this is *not* a property of the interacting cluster problem (see O'Kelly, 1992). A solution to this partitioning problem has recently been proposed by O'Kelly (1992) and this section briefly summarizes the method.

It is required to cluster the n observations into p groups, so that the sum of squared deviations from the cluster means is as small as possible. Assume that the cluster means are adjusted to reflect the interaction between the entities. Further, since it is desirable to place highly interactive observations in the same group, it will be assumed that the penalty for assigning observation i to group g and observation j to group h is an increasing function of the distance between the group centroids and the interaction level W_{ij}. Specifically, the 'cost' of assigning i to g and j to h is:

$$P[i(g), j(h)] = W_{ij}(d_{ig} + d'_{gh} + d_{hj}) \qquad (1)$$

where
W_{ij} is the exogenous interaction effect,
d_{ig} is the squared distance from i to cluster centre g,
d'_{gh} is the squared distance between the cluster centres,
d_{hj} is the squared distance from j to cluster centre h.

Let (X_g, Y_g) be the centroid of group g for $g = 1, \ldots, p$. The distances are defined as: $d_{ig} = (x_i - X_g)^2 + (y_i - Y_g)^2$ for all $i = 1, \ldots, n$ and $g = 1, \ldots, p$; and $d'_{gh} = (X_g - X_h)^2 + (Y_g - Y_h)^2$ for all g and $h = 1, \ldots, p$. In the first part of this analysis the objective is to choose (X_g, Y_g), $g = 1, \ldots, p$, so as to

$$\text{MIN } T = \Sigma_i \Sigma_j W_{ij} \Sigma_g \Sigma_h K_{ijgh} D_{ijgh} \qquad (2)$$

where $K_{ijgh} = 1$ if i belongs to group g and j belongs to group h, and $K_{ijgh} = 0$ otherwise; and where $D_{ijgh} = d_{ig} + d'_{gh} + d_{hj}$. Note that the K_{ijgh} integer variables obey the following restrictions:

$$K_{ijgh} = X_{ig} X_{jh} \qquad (3)$$

where the X values are allocation variables, that is:

$$\Sigma_j X_{ij} = 1, \text{ for all } i \text{ and } X_{ij} \text{ is either 0 or 1} \qquad (4)$$

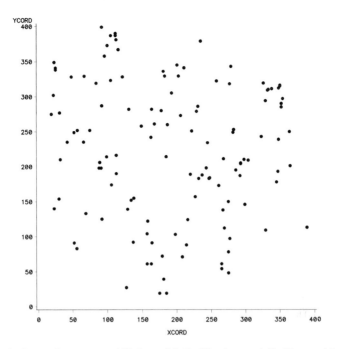

Figure 4.1. Point data set n = 125 from M. R. Klauber and P. Mustacchi (1970).

A useful property of the problem is that the use of a squared distance term yields a linear system of equations for the coordinates of the cluster centroids. These equations are derived and solved repeatedly, for a given set of cluster allocations. A sequential reallocation of the observations between the clusters is then performed: that is K_{ijgh} is initially assumed to be fixed and this is equivalent to starting the problem with a known partition of the observations between p groups. The model solved in O'Kelly (1992) discusses the iterative reassignment of observations to clusters.

As an example of the procedures explained in the previous paragraphs, consider the 125 points shown in Figure 4.1. The observations are used because they present a convenient source for a set of x, y locations and a time stamp for each event. No substantive contribution to the original data context is attempted here; rather suppose that these are the locations of fires in a city, and we are interested to see if there are clusters of events in space, in time, or in space and time. For the sake of illustration, a simple set of interactions between entities is modeled as $W_{ij} = 10./(EPS + | t_i - t_j |)$ where EPS is a small constant (set to 0.1) to prevent division by zero if the two events occur at exactly the same time. The result of clustering 125 observations into 4 groups is shown in Figure 4.2 which shows conventional group centroids, using an 'X' symbol, and the cluster membership of the data points. No attempt has been made here to find the optimal number of groups, and it is

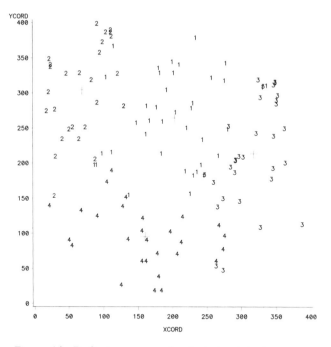

Figure 4.2. Preliminary example: clustering into 4 groups.

recognized that many different partitions of the data could be produced by altering the initial partition, or by changing convergence criteria.

While conventional clustering has been in use for many years as a means of spatial pattern analysis (see Baxter, 1971, for early examples) the extension of clustering techniques to allow for interdependence between the clusters adds significant benefits in GIS applications such as pattern recognition – e.g. hypothesis testing for patterns of arson and crime. There exists some strong potential applications of these tools, e.g. K-means program/ISODATA package; similar problems arise in the case of organizing large operations like the postal service, mail order deliveries and retail market delimitation.

Both the OR questions (concerning the optimal partitioning) and the GIS question (concerning the display/visualization of large data bases) require further work.

Lines and flows: spatial interaction

Consider the location of facilities, using an interaction model to drive the choice of sites. The probabilistic allocation of demand to these facilities is a classic problem in spatial interaction theory. Techniques to locate these facilities in a manner which optimizes the spatial interaction properties are well known (see Hodgson, 1978; and O'Kelly, 1987.) However, a new feature which is essential to these models, was omitted until the graphic capabilities of GIS revealed its importance.

Spatial interaction concepts are linked to covering and optimal location models in the following way: it is apparent from the visualization tools that the role of spatial constraint in interaction models has not been fully integrated into locational analysis and indeed the interplay with the space-time prism model is a very interesting possibility. If a facility has a known range, and therefore can only cover a demand point in the fixed radius, then this fact ought to be built into the spatial location model. The hybrid concept of probabilistic interaction towards facilities that have a covering radius introduces some subtle new spatial situations. Observe that there is no guarantee that a specific demand point has any facility within in its range. The demand from that origin is therefore uncovered, but in spatial interaction terms some demand must be sent to an artificial destination to represent the uncovered demand.

To pull out the general point: spatial analysts are concerned to link methods from OR etc., and to develop novel visualization tools to help with analysis of the model. The algorithm does the work, and then the solutions are posted back to the GIS to validate and query the quality of the solution. Questions are clear: are the correct assumptions built into the optimization tool? Are the key parameters and mechanisms represented in the model? What data/empirical estimates need to be adjusted as a result of the analysis? The entire model-building exercise is viewed as an iterative one that cycles back to the beginning to check the validity of the model – a step which is too often replaced by hand waving.

The problem is a generic one – whenever a complex model produces output in the domain of X, the sensitivities of outputs to changes in the key parameters, or to changes in the assumptions about the role of critical variables, must be investigated. The answers in this kind of research hinge on both a fundamental substantive understanding of the problem, and a mathematical insight to the model.

Spatial autocorrelation

As another example of the marriage of spatial models and GIS, consider the example of spatial autocorrelation. The correct measurement of spatial autocorrelation is a necessity, but one which is open to a wide variety of subtle variation. Clearly GIS allows these measures to be gathered rather easily, since adjacency is one of the properties of spatial entities that is maintained in GIS databases. This is possible because of the increased availability of adjacency 'facts' from topologically integrated databases. Even with spatial autocorrelation at hand, the analyst must decide how to make appropriate statistical use of the indices, and be cognizant of the role of units and scale of analysis on the results (Chou, 1991; Newsome, 1992). Also, the theory underlying the appropriate statistical operators for various measures of spatial autocorrelation needs to be fully understood by the end user (see, for example, Haining, 1978). There is little to be gained by making spatial autocorrelation one of the many descriptive statistics collected from a spatial database, unless the sophistication of the user is sufficient to make correct use of this information (see Getis, 1991; Chou, 1991).

Spatial situation vs. spatial site

A final example grows from the theme developed by Couclelis (1991) about the importance of site and situation and is discussed in terms of Fotheringham's competing destination (CD) models, and the possibility of enhanced measurement of CD effects in GIS.

An example of the kind of situation effect which is widely known is the Fotheringham (1983) measure of competition between destinations. The measure is important to spatial interaction models and can be expected to be an easily calculated variable in GIS implementations of spatial interaction models. Moreover, the GIS focus allows us to see that the conventional CD measure can be generalized once the ideas of neighbourhood variables and adjacency are fully integrated.

The starting point is $A_{ij} = \Sigma_{k \neq j, \ k \neq i} W_{ik} f_i(d_{jk})$, which is the Fotheringham measure of CD for facility j, from i's perspective. There are three reasons that the accessibility of j is measured with an 'i' subscript: (1) the calculation avoids the inclusion of i; (2) the attraction variable W is possibly calculated differently from each zonal perspective; and (3) the distance decay function is specific to zone i. Notice that this calculation includes all potential competitors of j, including some that might reasonably be expected to avoid interaction with i. The following set of generalizations emerge, and these are measures which would be easy to calculate in a database that keeps track of the 'neighbourhood' variables. Some care would have to be taken to ensure that the data structure allowed the efficient collection of the facts needed for the measures below. Two examples of each of three types of generalization are given here:

1 Restrictions on the list of competitors of j, from i's perspective

(a) $A_{ij} = \Sigma_{k \neq j, \ k \in S(i)} W_{ik} f_i(d_{jk})$, where $S(i) = \{k| \ d_{ik} \geq L\}$

This measures the accessibility of j to those facilities 'k' which are further than L from i (e.g. consistent use of the masking radius used in Fotheringham, would set L = 160).

(b) $A_{ij} = \Sigma_{k \neq j, \ k \in T(i)} W_{ik} f_i(d_{jk})$, where $T(i) = \{k| \ d_{ik} \leq d_{ij}\}$

This measures the accessibility of j to those facilities 'k' which are closer than d_{ij} to i (e.g. it could also be defined with inequality reversed).

2 Restrictions on the list of competitors of j, from j's perspective

(a) $A_{ij} = \Sigma_{k \neq j, keS(j)} \, W_{ik} \, f_i(d_{jk})$, where $S(j) = \{k | \, d_{jk} \leq Z\}$

This measures the accessibility of j to those facilities 'k' which are closer than Z to j (e.g. this is called a traffic shadow effect, in Taaffe, 1956).

(b) $A_{ij} = \Sigma_{k \neq j, keT(j)} \, W_{ik} \, f_i(d_{jk})$, where $T(j) = \{k | \, D_{jk} = 1\}$

This measures the accessibility of j to those facilities 'k' which are in the set of pairs for which $D_{jk} = 1$. These could be pairs of twin cities, or other close substitutes.

3 Restrictions on the list of competitors of j, from (i, j)'s perspective

(a) $A_{ij} = \Sigma_{k \neq j, keS(i,j)} \, W_{ik} \, f_i(d_{jk})$, where $S(i, j) = \{k | \, d_{ij} + d_{kj} \leq R\}$

This measures the accessibility of j to those facilities 'k' which are in the set of pairs (j, k) for which the sum of distances from i to j and from k to j is less than some budget. This could be used to define an interaction space such as a corridor.

(b) $A_{ij} = \Sigma_{k \neq j, keT(i,j)} \, W_{ik} \, f_i(d_{jk})$, where $T(i, j) = \{k | \, d_{ik} + d_{kj} \leq R_1$, $d_{ik} + d_{kf} \leq R_2, \, d_{jk} + d_{kf} \leq R_3\}$

This measures the accessibility of j to those facilities 'k' which are in the action space bounded by distances of R_1, R_2, and R_3 from the fixed points i, j, k and f. These could be pairs of interactive points, or corridors pointing towards some particularly attractive alternative such as the city centre 'f'.

Important research questions could then be answered in spatial analysis: such as further empirical evidence of the role of the competition effect, especially in the light of the generalized measurement of spatial competition. With these tools realism is added to the theoretically justified measures of competition between alternatives proposed by Fotheringham (1983).

Progress via incremental improvements

There are many spatial analytical tools which do not currently exist in GIS (and some which may be inherently impossible to implement because of data structures – see Couclelis, 1991; Goodchild, 1987). In addition GIS prompts us to see new ideas in spatial analysis, which otherwise might not be clear. But there are barriers to the realization of this potential.

To get from the current level to some of the tools discussed here, major fundamental research will have to be done on some basic steps. These intermediate steps include:

1. Exploratory data analysis, to search for previously unseen processes within the data and test new analytical techniques for finding new instances of a sample pattern.

2. Systematic integration of existing models of spatial processes such as interaction and gravitational concepts, spatial autocorrelation measures etc. This includes investigation of the problem of 'modifiable areal units'.
3. Spatial process and spatial patterns through time also need to be investigated.

The advanced tools discussed in this chapter must not be viewed as modules to be added 'piecemeal' to the functionality of a GIS: rather these are themes which underlie the development of a complex set of new procedures. These themes will surface in various ways and in differing proportions depending on the specific domain of research. What are the implications for spatial analysis? While there are many (see Openshaw, 1991; Nyerges, 1991; and others) just two issues are mentioned here. The first is that exploratory data analysis (EDA) will be the tool of choice; the second requires analysts to rethink the role of geometric spatial analysis. These issues are dealt with in turn.

Exploratory data analysis

Techniques for sifting through large data streams and innovative visualization techniques are needed to allow the analyst to assimilate the quantity of data presented (e.g. imaging systems; panel surveys; census products). The major problem becomes that of deciding what is important in the vast quantities of data which are generated in large models and GIS. The key technical advance will be in pattern recognition, which intelligently allows the user to sift through the data, reduce dimensionality, find patterns of interest, and then order the GIS to find other instances or similar occurrences. This sounds simple, but is difficult to implement when the size of the underlying data base is of the order of multiple gigabytes.

Also in the arena of *exploratory data analysis* and *categorical data analysis* is the important point that large-scale data surveys in the *social sciences* are producing very large amounts of data. The quantity of these data may discourage analysts from embarking on useful research. In some cases the quality of the data is also in question, especially in the marketing and retail analysis arena where the results of cluster analysis are used to estimate the micro demographics of spatial zones. Other high-quality examples which are currently available include the *Annual Housing Survey*, data collected from continuous work history, the Italian census as discussed by Openshaw, the Cardiff panel data survey, etc. In the presence of such large data sets the need for novel visualization techniques in the social sciences is just as pressing as in the physical realm.

Geometric pattern analysis

At the earliest stage of quantitative analysis, simple small-scale analyses were performed. Typically, these data analyses considered small sample case studies, and involved a proof of concept kind of approach. The full 'industrial strength' scale up of these methods was constrained by the power of data processing technology, and by the lack of large digital databases. Now, with the advent of digital databases, conventional spatial statistical studies using nearest neighbour analysis, quadrat analysis, or other spacing statistics (e.g. Getis and Boots, 1978; and Rogers, 1974) can be reinvigorated when matched to very large data sets, and to efficient computational geometry routines. A broader issue remains: are the types of questions and analyses carried out earlier still interesting? Are these techniques tractable for data sets of arbitrary size? Are heuristics needed? Do these breakthroughs allow us to give trustworthy answers (Openshaw, 1987; Openshaw *et al.*, 1987) to reasonable questions about spatial patterns? Researchers need to re-examine the tool kit of spacing, quadrat and nearest neighbour techniques, and assess the impact of larger data volumes and improved computational technique on their applicability.

While short-run questions challenge us, spatial scientists must also aim for the integration of spatial models into GIS and anticipate the successful marriage of theory and practice, yielding solutions to problems that were previously thought to be unmanageable.

Conclusion: barriers to improving GIS via spatial analysis tools

As a conclusion to this chapter, let us examine, briefly, the capability of GIS for spatial analysis tasks, and highlight some of the barriers to realizing the potentials mentioned in this paper.

One of the observations that is made repeatedly is that there is a mismatch between the spatial analytical capabilities of the research community and the applied tools available in GIS and in use by practitioners. It would seem to be obvious that GIS would be improved immediately if only all the research capabilities of the best spatial analysts were somehow 'built in' to GIS. This has not happened and the result is that exciting research tools are not quickly adopted by GIS and are inaccessible to the practitioner. Three reasons for the breakdown in transfer of technique to technology appear: first, because the non-geographers using GIS do not think in spatial terms, and therefore do not intuitively ask questions about spatial pattern and spatial association. For this group, space as a container is a perfectly acceptable medium, and the GIS as a spatial data handling tool provides all the functionality (see Couclelis, 1991). Second, the research community has not been able to justify the models using real world 'value added' terminology. Therefore, research tools remain in the laboratory, without being implemented in commercial packages. And third, the real world is driven by market concerns and perceives spatial research tools as limited by the size of the market (see similar point in Goodchild, 1987, p. 333).

For theoreticians the first priority is *not* that ideas be commercially applied, as the worth of a method is not judged solely in terms of its marketing in a commercial package. However, if geographers want to participate fully in this environment, *they have to be more concerned with explaining research results in a manner which is accessible and where the marginal value of an enhancement to a research tool is clear*. This may mean being careful not to leave the discussion in the hands of some less technical marketing person. Changes *have* taken place, and the realities point to a much more commercially oriented research environment than before. It is important that the people who were instrumental in nurturing this branch of the discipline learn the additional skills necessary to ensure the timely and accurate usage of sophisticated spatial tools. These commercial prospects also introduce heightened competition, ethical issues, and legal issues, and these are part of the costs associated with increased real world interest in the realm of spatial analysis. As a side-benefit, the GIS explosion has ended the debate about 'relevance' of quantitative methods, and the growth in GIS enrolment in graduate programs has reinvigorated the quantitative courses that go along with this track; it is instructive to reassess some of the essays in Gaile and Willmott (1984), and in Macmillan (1989) in the light of the fast pace of GIS growth. Openshaw (1991) puts this point of view in especially clear terms.

Taking stock, there is a lot to be happy about – the reinvigorated role of spatial analysis, and the increased visibility of the research workers at the cutting-edge of this discipline. However, specialists in this area must consider the challenge of taking research tools out of the lab and into the marketplace for consumption by end-users.

Appendix

Throughout this chapter I have kept the discussion 'neutral' with respect to software and commercial packages. This set of notes adds specific names and products in parenthetical form.

1. Examples on the cutting edge include McDonald's use of GIS platforms to maintain customer spotting data, and the portrayal of multi-media reports on retail sites. Nevertheless, it is not surprising that the majority of commercial real estate research departments are using simple demographic desk top mapping packages, and are a long distance away from adopting optimal centralized site selection. Even in redistricting software, there is an emphasis on mapping impacts of boundary changes, driven by the analyst's judgement, rather than using algorithms to define optimal partitions. In any business there is a spectrum of innovativeness all the way from visionary advanced guards (e.g. McDonald's and Arby's) to relatively staid followers, who are not interested in innovation. There is a market for spatial analysis, but at the moment it is directed towards simple

rather than complex spatial data handling tasks. The market will grow more quickly as the adoption of more sophisticated ideas penetrates the user community, and trained spatial analysts will be called on to explain and implement technical advances.

2. The SPANS package allows the demand from a set of areas to be allocated among a fixed set of facility locations.

3. Many associate GIS with the commercial software packages that are on the market today. These packages often provide elementary analytical tools, but quite often they stop short of the full flexibility of the individual's expertise. That is, individuals writing their own personalized code would add features that would not have broad appeal, or indeed broad acceptance and understanding. Thus there is a problem – it is not possible to tackle cutting-edge research problems solely within the confines of commercial GIS, because by definition they only include the more widely accepted and known tools. The exception would be that the GIS may provided a 'macro', 'procedure' or 'application' writing capability, through which the skilled analyst can build a complex model from fundamental building blocks. Some packages have such diverse tools that it is possible to create macros for novel spatial analytical techniques.

Acknowledgements

The support of the National Science Foundation, SES 88-21227 and SES 89-46881 for 'Models of the Location of Hub Facilities' is gratefully acknowledged. NSF Grant SES 88-10917 supported participation in the NCGIA I–14 initiative. Support from the Ohio Supercomputer Center for 'A clustering approach to the planar hub location problem' is also appreciated. Research assistance has been provided by Harvey Miller, and his comments on the manuscript are appreciated. An earlier version of this chapter was presented at San Diego State University, 14 February 1991, where useful comments were received from Stuart Aitken, Serge Rey, Gerry Rushton and Art Getis. Discussions at the I–14 meetings with Howard Slavin and Noel Cressie were valuable. Comments have also been received from the Geodetic Science seminar at The Ohio State University.

References

Bartlett, M. S., 1975, *The Statistical Analysis of Spatial Pattern*, London: Chapman and Hall; New York: Wiley.

Baxter, R. S., 1971, *An urban atlas – Reading*, prepared within the Urban Systems Study under the sponsorship of the Centre for Environmental Studies. Cambridge, University of Cambridge Department of Architecture.

Chou, Y. H., 1991, Map resolution and spatial autocorrelation. *Geographical Analysis*, **21**(3), 228–246.

Couclelis, H., 1991, Requirements for planning-relevant GIS: a spatial perspective, *Papers in Regional Science*, **70**(1), 9–19.

Diggle, P., 1983, Statistical Analysis of Spatial Point Patterns (Mathematics in biology), London: Academic Press.

Fotheringham, A. S., 1983, A new set of spatial interaction models: the theory of competing destinations, *Environment and Planning A*, **15**, 15–36.

Gaile, G. L. and Cort, J. W. (Eds.), 1984, *Spatial Statistics and Models*, Dordrecht; Boston: D. Reidel Pub. Co.; Hingham, MA: distributed in the USA by Kluwer Boston Academic Publishers.

Gatrell, A. C., 1983, *Distance and Space: a Geographical Perspective*, Oxford: Clarendon Press; New York: Oxford University Press.

Getis, A. and Boots, B., 1978, Models of Spatial Processes: an Approach to the Study of Point, Line, and Area Patterns. Cambridge (Eng.); New York: Cambridge University Press.

Getis, A., 1991, Spatial interaction and spatial autocorrelation: a cross-product approach, *Environment and Planning A*, **23**, 1269–1277.

Glick, B., 1979, Tests for space-time clustering used in cancer research, *Geographical Analysis*, **11**(2), 202–207.

Goodchild, M. F., 1987, A spatial analytical perspective on geographical information systems, *Int. J. Geographical Information Systems*, **1**(4), 327–334.

Goodchild, M. F. and Janelle, D., 1984, The city around the clock: space-time patterns of urban ecological structure, *Environment and Planning A*, **16**, 807–820.

Haining, R. P., 1978, Specification and estimation problems in models of spatial dependence. Evanston, Ill.: Dept. of Geography, Northwestern University.

Hodgson, M. J., 1978, Towards more realistic allocation in location-allocation models: an interaction approach, *Environment and Planning A*, **10**, 1273–1285.

Janelle, D., Goodchild, M. F. and Klinkenberg, B., 1988, Space-time diaries and travel characteristics for different levels of respondent aggregation, *Environment and Planning A*, **20**, 891–906.

Kent, S., 1984, Analyzing activity areas: an ethnoarchaeological study of the use of space, Albuquerque: University of New Mexico Press.

Klauber, M. R. and Mustacchi, P., 1970, Space-time clustering of childhood Leukemia in San Francisco. *Cancer Research*, **30**, 1969–1973.

Knox, E. G., 1964, The detection of space-time interactions. *Applied Statistics*, **13**, 25–30.

Macmillan, Bill, 1989 (Ed.), *Remodelling Geography*, Blackwell: Oxford.

Miller, H., 1991, Modeling accessibility using space-time prism concepts within geographical information systems, *Int. J. of Geographical Information Systems*, **5**(3), 287–301.

Newsome, T. H., 1992, *Measuring Spatial Pattern in Census Units: Residential Segregation in Franklin County, Ohio*, Unpublished Ph.D. Dissertation, Department of Geography, The Ohio State University, Columbus, Ohio.

Nyerges, T., 1991, Geographic information abstractions: conceptual clarity for geographic modeling, *Environment and Planning A*, **23**, 1483–1499.

O'Kelly, M. E., 1987, Spatial interaction based location-allocation models, *Spatial Analysis and Location-Allocation Models*, A. Ghosh and G. Rushton (Eds.), New York: van Nostrand Reinhold.

O'Kelly, M. E., 1992, A clustering approach to the planar hub location problem, *Annals of Operations Research*, **40**, 339–53.

Openshaw, S., 1987, An automated geographical analysis system. *Environment and Planning A*, **19**, 431.

Openshaw, S., Charlton, M., Wymer, C. and Craft, A., 1987, A Mark I Geographical Analysis Machine for the automated analysis of point data sets, *Int. J. Geographical Information Systems*, **1**(4), 335–358.

Openshaw, S., 1991, Commentary: A view on the GIS crisis in geography, *Environment and Planning*, **23**, 621–628.

Parkes, D. and Thrift, N., 1980, Times, spaces, and places: a chronogeographic perspective. Chichester (Eng.); New York: John Wiley.

Rogers, A., 1974, Statistical analysis of spatial dispersion: the quadrat method. London: Pion; New York: Distributed by Academic Press.

Taaffe, E. J., 1956, Air transportation and United States urban distribution, *The Geographical Review*, **46**(2), 219–238.

Upham, S., 1979 (Ed.), Computer graphics in archaeology: statistical cartographic applications to spatial analysis in archaeological contexts. Tempe: Arizona State University.

Villoria, O. G., 1989, *An Operational Measure of Individual Accessibility for use in the Study of Travel-Activity Patterns*, Unpublished Ph.D. Dissertation Department of Civil Engineering, Ohio State University.

PART II

Methods of spatial analysis and linkages to GIS

5

Two exploratory space-time-attribute pattern analysers relevant to GIS

Stan Openshaw

Introduction

There is seemingly much confusion about the nature of spatial analysis considered relevant to GIS databases. However, it is argued that the GIS revolution and the increasing availability of GIS databases emphasizes the need for what are primarily exploratory rather than confirmatory methods of spatial analysis and computer modelling (Openshaw, 1989). Techniques are wanted that are able to hunt out what might be considered to be localised pattern or 'database anomalies' in geographically referenced data but without being told either 'where' to look or 'what' to look for, or 'when' to look. The increasing availability of georeferenced data emphasizes the need for a new style of data-driven, maybe even artistic and insightful, exploratory forms of analysis that greatly reduces the traditional emphasis on inferential methods. Additionally, whether researchers like it or not, the spatial analysis task within GIS is starting to be driven by the availability of locationally referenced data that suddenly becomes available in a form suitable for analysis. Indeed, there is an increasingly strong imperative to analyse data purely because they are now available for analysis, despite the absence of either an a priori experimental design or testable hypotheses. This is quite different from how spatial analysis or spatial statistical methods have tended to be applied in the past.

This chapter seeks to develop GIS analysis tools able to explore or trawl or screen GIS databases for evidence of pattern that exist in the form of data anomalies of various kinds. The difference between a database anomaly and pattern is partly a matter of semantics but the distinction is useful because it emphasizes that such anomalies need not be 'real' and the analysis process can only identify anomalies as potential patterns and then pass the task of validation to other technology [see Besag et al. (1991) and Openshaw and Craft (1991)]. This is clearly only one form of spatial analysis. The interesting new suggestion here is that this pattern spotting technology should not just work in geographic space but operate simultaneously in the other two principal data domains that typify many GIS databases: namely temporal space and more generally in the attribute space. This tri-space analysis capability is seen as being very important because it opens up the totality of the GIS environment to pattern spotting procedures. Various tools exist for exploring and analysing these three spaces separately and a few methods can cope with two of the three spaces, for instance, space-time statistics. None so far has appeared that can handle all three simultaneously or any permutation of the three data spaces. Likewise, the visualization of spatial information is basically restricted to two of the spaces and to relatively low levels of data dimensionality. This absence of effective tri-space analysis tools is potentially a severe problem in a GIS environment because ideally the spatial analysis function needs to cover all three types of spatial information. This is particularly important in an exploratory context where the whole philosophy is to 'let the data speak for themselves' and suggest, with a minimum of pre-conditioning, what patterns might exist. Indeed, without this extended functionality it may never be possible to extract maximum value out of GIS databases.

This chapter is concerned, therefore, with developing a prototype analysis procedure that is able to explore tri-spaces for pattern. The chapter tackles some of these issues by outlining the development of a new and more flexible approach to pattern spotting that is relevant to GIS. Sections 2 and 3 describe some of the basic design criterion that would appear to be important in attempting to meet the goals of exploratory spatial data analysis within a GIS environment. Sections 4 and 5 describe two alternative procedures. Section 6 briefly describes their application to a GIS database and Section 7 offers some more broadly based conclusions as how best the technology might be further developed.

Exploring GIS databases

In seeking to explore a GIS database, there are a number of key assumptions that need to be made. In particular, it is assumed that the GIS contains space-time-attribute patterns that are:

1. unknown and sufficiently important for their discovery to be useful and insightful;
2. real and meaningful in relation to the data being examined;
3. in a form that can be found by a highly automated computer based tool of the type described in Sections 4 and 5;
4. and users who are: sufficiently interested in analysis, or the possibility of analysis, to make the development of such pattern spotters worthwhile; and
5. have a degree of faith that exploratory pattern analysis is both possible and likely to be sufficiently useful to justify the effort involved in developing the necessary technology.

The principal difficulty in this as well as in other areas of spatial analysis is that most end-users and vendors still seemingly do not properly appreciate the need for sophisticated analysis tools. This reflects many factors, particularly: a lack of appreciation of what might be possible, the absence of spatial analysis traditions, a lack of market-place demand, and a concentration on the database creation aspects of GIS; see also Openshaw (1989, 1990a, 1991). On the other hand, it is apparent that many users may be facing a potential dilemma. Once their databases are complete, they will be under increasing pressure to analyse data that was collected for other purposes. It is quite natural that they will expect GIS to assist them in this task. It is likely that many classes of GIS users will one day, probably fairly soon, need access to pattern spotting data exploration tools. Existing spatial statistical methods and quantitative geographic techniques are generally not adequate for this task, see Openshaw and Brunsdon (1991). A good example is spatial disease epidemiology where the availability of georeferenced data has stimulated an immense amount of activity of precisely this nature but based mainly on inadequate analytical technology; see for example, Openshaw and Craft (1991), or Besag *et al.* (1991). More generally, there is an obvious need to be able to analyse the emerging heap of geodatabases created by GIS. The necessary tools have to be clever enough to filter out the rubbish, to ignore data redundancy, overlook the insignificant, and yet powerful enough to identify possible pattern against a background of data uncertainty and noise. A new breed of database analysers are needed that are able to seek out systematically patterns that are in some ways unusual amongst masses of spatial information, some of which is probably meaningful and much of which is potentially junk.

Aspects of designing a tri-space data analysis tool

3.1. Understanding the problem

This exploratory analysis task is not easy and can be characterized by a number of general characteristics.

1. The data to be analysed are often determined by circumstances that are initially, and possibly always, beyond the analysts' control.
2. There is often little or no prior knowledge of what patterns might be expected or any relevant hypotheses to test. Indeed, an important objective of the analysis is to generate new hypotheses or uncover insights about patterns and relationships that may exist in the data.
3. The available information will typically be non-ideal. It may not be completely accurate or free from spatial bias. Additionally, false patterns may well exist either because of spatially structured differences in data reliability or because of error propagation within GIS or because of a mixing of different data sources from different spatial scales and/or time periods, or because of other unnoticed data error. GIS databases constitute a noisy and non-ideal data environment and the analysis tools have to be able to cope with this situation.

4. The data are often strongly multivariate and contain a mix of useful and useless variables but without any obvious or easy way of selecting the most appropriate ones. Additionally, much of the information may well be in surrogate form providing poor proxies for other more relevant but missing variables.
5. Data volumes may be large.
6. Typically there will be a mix of all possible measurement scales.
7. There is often an urgent imperative placed on the analysis task that insists that the database should be analysed even though it is non-ideal. The attitude of 'let's see what we can do with what we have got' is becoming increasingly prevalent.
8. It is important to avoid an overly simplistic approach. There is no reason to assume a priori that large scale spatial heterogeneities exist or that patterns occur in many locations rather than just a few. Instead of imposing possibly inappropriate theoretical pattern structure on the data, why not let the data themselves define what structures are relevant and where they may exist? This requires analysis tools that are both highly flexible and data context adaptive with a minimum of preconditioning as to what is and what might not be important.
9. There is some need for a high degree of analysis automation in order to cope with the large and rapidly growing number of data sources that potentially require spatial analysis.
10. There is an increasing need for what might be termed 'real-time' spatial data screening on databases that are seconds or minutes or hours or days old, rather than years or decades out of date.

Some users have an urgent need for exploratory analysis tools that can operate sufficiently fast to allow a proactive rather than a retrospective response and this type of application is likely to become extremely important. Another important problem is that the results are dependent on a number of arbitrary operational decisions concerning the selection of data to analyse. Traditional science emphasizes the importance of a priori selection but this is hard in a GIS context in which the analysis task is essentially out of control and unguided. Nevertheless, a number of key operational decisions can be identified that may well critically affect the results of any analysis. They include:

1. choice of study region;
2. selection of data to study; for example, choice of disease and other forms of data disaggregation;
3. choice of covariates that are to be controlled for;
4. selection of scale, aggregation, and choice of areal units;
5. choice of time period or periods for analysis; and
6. (if relevant) the choice of hypotheses to test.

Some of these decisions will be informed by purpose. Whilst most are essentially arbitrary, they need be neutral in relation to their impact on the results of the analysis. For example, a time varying process may well be extremely sensitive to the choice of time periods for analysis.

Quite often these data selection decisions reflect the pragmatic desire to manipulate the data into a form suitable for analysis and the analyst seldom seems to worry unduly about the impact of such decisions on the end results. At the very least some sensitivity analysis should be performed but this is seldom, if ever, done in practice. In a GIS context the analyst is forced to make many selections prior to starting spatial analysis; for example, in spatial epidemiology, the choice of disease types (and thus cases) to study, the selection of one or more time periods, the choice of population at-risk-data, and selection of study region. Additionally, it appears that probably the only patterns that can be found at present are those that exist in a purely spatial domain. Yet even these results are conditioned by all those user decisions that have themselves been superimposed on top of many others that were involved in the creation of the database in the first place. It is argued that this is a potentially very dangerous situation to be in and that more flexible spatial analysis tools are needed if best use is to be made of many GIS databases in an exploratory context.

Using all the data with a minimum of pre-selection

A desirable goal should be the development of tools that are able to minimize the number and level of subjectivity involved in the exploratory data analysis task. A worthwhile general principle should be to develop methods of analysis that impose as few as possible additional, artificial, and arbitrary selections on the data. In short the data should be analyzed at the finest resolution at which they are available, in their entirety, and this principle should be applied to all aspects of the database and all forms of exploratory analysis. Computers are now sufficiently fast such that there is no longer any requirement for the analyst to impose a large number of arbitrary data selection decisions purely so that the analysis task remains tractable. New computer analysis tools are needed that can exploit the availability of this computational power by exploring GIS databases in their entirety using all available information. It should be an important function of the analysis technology (and not the analyst) to discover how best to subset the data. Or at the very least, the analyst should have the option not to make possibly 'key' choices if he does not wish to.

Exploring the three principal GIS data spaces

It follows then that analysis technology is needed to search for patterns by operating simultaneously in all the relevant data spaces found in a typical GIS database. Indeed it is worth emphasizing that a GIS database usually consists of three principal types of information:

1. geographic or locational information;
2. temporal details; and
3. other attribute data.

Most GISs only distinguish between (1), and (2) plus (3) together. Each of
these information types represent data spaces that are characterized by different measurement metrics. Geographic space is bi- (or tri-) dimensional,
and can be regarded as continuous. Temporal space is usually one dimensional, and continuous. The other attribute data are typically multivariate
with a mix of measurement metrics that are seldom continuous. The three
spaces do not share a common metric nor can they easily be transformed into
a comparable form. Moreover, all three spaces can be fuzzy. Traditionally,
each data space has been analyzed separately but there is clearly a need to
discover how best to analyse them together.

The GAM-type of spatial pattern spotter only works in geographic space
with an arbitrary choice of what to analyse being based on subjective decisions regarding the other two spaces; see Openshaw *et al.* (1987, 1990b) or
Besag and Newell (1991). It is suggested that the analysis of purely spatial
patterns, that have probably been created by massive subjectivity regarding
time and attribute spaces, are only ever likely to be of limited interest. It
follows that laudable attempts, such as that by Diggle (1991), to develop
more refined process orientated tools for purely spatial analysis are unnecessary because of this fundamental limitation. Time-series analysis methods
handle purely temporal data with selections made in terms of geographic
space (viz. choice of study region) and attribute space (viz. choice of variables).
Attribute-only analysis involving classification or other multivariate statistical
methods is based on selections made in geographic space, in temporal space,
and also often in attribute space. In theory, different combinations of these
spaces might be investigated. Table 5.1 lists the various data space permutations. However, to date only space-time interactions seem to have been considered; see for example, Knox (1964), and Mantel (1967). Seemingly, these
methods have advanced little in the last 30 years; see Marshall (1991) for a
review.

It is argued that the GIS era emphasizes the need for a new generation
of spatial data exploratory technology that is able to operate simultaneously

Table 5.1. Different data space interactions.

1. geographic space only
2. geographic space – time space
3. geographic space – attribute space
4. geographic space – attribute space – time space
5. time space only
6. time space – attribute space
7. attribute space

Note: the ordering is irrelevant

in all three data spaces, exploring all the information likely to be found in a database, without much (or any) preselection. New analysis methods are needed that do not have to be told where to look for pattern or when to look for it or what attributes are responsible for the patterning. At the same time, these exploratory tools are only screening or trawling devices. They are meant to identify possible patterns in the form of database anomalies. They cannot validate their existence in any formal inferential sense but only suggest where to look and where further research might be needed. The results are designed to be suggestive rather than definitive.

So what is pattern?

Some notion of 'pattern' is still needed. In a GIS, pattern may often be viewed as a localised excess concentration of data cases that are unusual, and thus of potential interest, either because of the intensity of their localised concentration or because of their predictability over time or their similarity in terms of their features. The overall assumption is of a search for localized areas which are surprising or anomalous in some ways. There is no need to be scale specific. A flexible pattern spotter should be able to determine whether the patterns are highly localized, small or large, or generally recurrent over large areas. This is quite different from the normal statistical approach in which pattern may be hypothesized as existing at, for example, a regional level prior to any analysis. In an exploratory context, the analyst should not be making scale specific pattern assumptions that are then imposed on the analysis process from a position of ignorance. It is far better if the scale and nature of any pattern can emerge from the analysis rather than be imposed upon it. Indeed, the GAMs are already fairly successful at achieving this objective albeit in a purely spatial context.

A high degree of automation is desirable

A further design decision is the degree of analysis automation that is to be attempted. Traditionally, spatial statistical analysis has been an essentially manual process, usually within the context of some research design and inferential framework. Likewise, the statistical modelling of pattern generating processes is still largely viewed as a manual or interactive man-machine activity of the GLIM sort. This is adequate for smallish data sets containing few variables, obtained by sampling in some standard way. It is, however, not relevant to massively multivariate GIS databases with tri-space information based on non-sample spatial data. The human-being based eye-balling approach just cannot cope with the complexity of the task and for this reason traditional mapping, Tukey-style exploratory data analysis, and other visualization approaches will always be limited in what they can offer as raw data pattern spotters relevant to the GIS era. The use of human-being intelligence is not excluded but it needs to be saved for that part of the analysis task where these unique skills and abilities are most likely to be most useful. By all means, manually probe a GIS database by mapping key variables but unless the patterns to be spotted are so obvious that no clever or sophisticated analysis is needed, then this will not usually be sufficiently powerful to progress the analysis very far. It is argued that fully automated methods of data screening for pattern are needed in order to perform efficient, comprehensive, and detailed exploration of large and complex databases, methods that are able to leave no stone unturned and offering a minimum risk of missing anything important. There should also be some capacity for real time operation, so that databases might continually be analyzed and re-analyzed for pattern as and when new records are added to it.

Incorporating human knowledge is important

Traditionally, most aspects of the analysis process were human-being based. The computer was used essentially as an arithmetic device. Here the design strategy is quite different with an emphasis on minimizing the human inputs by deliberately seeking to adopt a cruder and totally computer intensive analysis approach. This leaves unresolved how best to incorporate human knowledge, intuition, and understanding into the analysis function. Automating the search process and the adoption of a look everywhere search strategy avoids any strong emphasis on hypothesis testing and it also avoids the weakness of the human-being being employed as a search tool. This is important as the amount of analyzable information continues to increase and avoids the dangers of overload and the likely absence of prior knowledge of what to look for. However, it should be recognised that automated analysis machines can only ever go a short way down the analysis route.

There are limits to what even intelligent machines are capable of doing. In the present context the claim to intelligence is fairly minimal and results from merely wishing to search everywhere for evidence of pattern. There is no mechanism for interpreting the results built into the search process and it is here where the human-being is best employed. The secret at present must be some kind of 'mix and match' strategy in which both the analysis machine and the analyst concentrate on doing what each is best at. Intelligent knowledge based systems (IKBS) and particularly neurocomputing will gradually reduce the amount of informed speculation and experience based interpretation that is still left to the analyst to perform. However, at present these systems do not exist and the human being is left with the still difficult task of making sense of what the pattern spotters may think are worthy of further investigation. It is matter for subsequent debate as to whether the pattern spotter is fine tuned to minimize what may still be termed type I errors to ease the analyst's interpretational load by deliberately only looking for the most massive evidence of pattern, or whether it is better to accept a higher type I error rate purely to avoid missing anything important.

A space-time-attribute analysis machine (STAM)

Origins

The STAM is one example of what might be achieved by the large-scale automation of a fairly simple analysis method. By searching all locations (in geography, time and attributes spaces) it is possible to overcome our ignorance of not knowing where to look. The motivation for this development was the analysis of leukaemia data. In the United Kingdom a national UK cancer case database exists which is geographically referenced to about 100 m resolution, the temporal resolution is measured in days, and a number of case attributes are also available but are expressed in a variety of different measurement scales (viz. age, sex and disease classification codes); see Draper (1991). Typically, a conventional space-only analysis of these data would involve an aggregation to census enumeration districts (a 100-fold coarsening of geographic space), aggregation into 20-year or 10-year time periods (a 365- to 730-fold reduction in temporal resolution), the selection of a single disease code and age-sex group (a 210-fold reduction) and the use of census population-at-risk data which is updated only once every ten years (a major source of errors and uncertainty). Indeed, the analysis of important rare disease data may be severely affected by the forced use of temporally incorrect census data and by fuzziness in the disease categorisations that are never reflected in the final results. It would be best to analyse this typically generic GIS database at the finest level of resolution and in its entirety to avoid making unwarranted assumptions.

Searching for space-time-attribute patterns via a brute-force approach

The basic philosophy is to change the emphasis from a manual, human-being, based spatial analysis process, to one in which the computer is used to perform all the hard work. It is simply assumed that computer power is in effect limitless and free. As Openshaw (1992) observes:

> 'Suddenly, the opportunity exists to take a giant step forward not by becoming clever in an analytical sense, but purely by becoming cruder and more computationally orientated in a way that allows the computer to do most of the work. It is almost as if computer power is to be used to make-up for our ignorance of the nature and behaviour of the complex systems under study'.

This same approach supported the development of Geographical Analysis Machines (Openshaw *et al.*, 1987; Openshaw, 1990b, 1991) and in some subsequent developments; for example, (Besag and Newell, 1991). Yet even a few years ago the idea of searching for spatial pattern without knowing either what to look for or where to look by considering all locations or particular sets of location, would have been both computationally unfeasible and totally alien. The latter difficulties still remain but the computational problems have disappeared. In less than three years, the early GAMs developed on a Cray X-MP supercomputer can now be run on a standard UNIX workstation. Subsequent developments of this search everywhere philosophy have resulted in a spatial relationship seeker known as a Geographical Correlates Exploration Machine (GCEM), see Openshaw et al. (1990) and in suggestions for the development of a Space-Time-Attribute Machine (STAM); see Openshaw et al. (1991). It is this latter technology which is further developed here.

The early GAM of Openshaw et al. (1987) briefly discussed the idea of extending the search for pattern from a purely spatial domain into time and attribute spaces. However, it was soon apparent that the original GAM search strategy of covering each space by a fine grid and then searching for pattern via a series of overlapping circles of varying radii would never be practicable with large data sets and tri-data spaces. For instance, whilst a typical GAM/1 run might well have involved several million search circles, an extended tri-space GAM would require several billion and current supercomputers are just not fast enough to cope.

A basic STAM/1 algorithm

The original idea in developing a Space-Time-Attribute Analysis Machine (STAM/1) is to search for pattern in the three critical data spaces using a search strategy based on observed data records. This at least localized the search in regions where there are data and thus some prospect of there being some pattern that might be detectable. The STAM/1 is essentially a fuzzy search process that looks for partial similarities in the three spaces. A basic algorithm is as follows.

Step 1. Define an initial range of search in the three data spaces; $g_1 \ldots g_n$ (for geographic space), $t_1 \ldots t_n$ (for time), and $a_1 \ldots a_n$ (for attributes). This is arbitrary but should not be too restrictive.

Step 2. Select an observed data record with a profile in g, t, and a spaces.

Step 3. Define a geographic search region centred on the observed case and distance radius g_r where $(g_1 < g_r < g_n)$.

Step 4. Define a temporal search region centred on the observed case and distance radius t_r where $(t_1 < t_r < t_n)$.

Step 5. Define an attribute search region centred on the observed case and distance radius a_r where $(a_1 < a_r < a_n)$.

Step 6. Scan through the database to find those records that lie within these critical regions. Abort search if fewer than a minimum number of records are found.

Step 7. Use a Monte Carlo significance test procedure to determine the probability of a test statistic for any search occurring under a null hypothesis.

Step 8. Keep the record identifier and the search parameters if the probability is sufficiently small.

Step 9. Examine all combinations of g, t, and a search parameters.

Step 10. Report the results if a sufficient number of the results pass the stated significance level.

Step 11. Change the central record, and repeat steps 3 to 11.

Some explanation might be useful. The search process is based on circles in all three spaces but each space has its own distance metric. The attribute data may actually require three or four different metrics rather than the one assumed here, with a corresponding increase in the number of search parameters and hence permutations to search. For example, in geographic space the search might range from 1 to 20 km. in 2 km. increments. Obviously the user has to apply a degree of knowledge and intelligence to select reasonable search parameters. The temporal search might be between 1 and 36 months in 3-month increments. The attribute search will dependent on the distance (dissimilarity) metric used but similar principles apply here also. Likewise it would be possible to experiment with relative rather than absolute measures of distance; for example, k-th nearest neighbours in each of the three data spaces.

In some ways the Step 7 Monte Carlo significance test procedure is the most critical. However, no formal test of hypothesis is intended and it is only used to screen out seemingly non-interesting results. A number of different test statistics could be used and maybe a Monte Carlo based test would not be required. For instance, if population-at-risk data are available then a Poisson probability could be used instead. Here it is assumed that no good population at risk data are available and a significance assessment is made on the relative ease of obtaining a similar or greater number of records that lie within the search spaces purely by chance. This is the usual null hypothesis generally used when nothing better is available.

The STAM ignores the multiple testing problem adopting similar arguments as used by Besag and Newell (1991); namely, that it is a database screening device and therefore is of little consequence. One possible problem is that the number of potentially interesting solutions found may well reflect the granularity of the search process. A useful additional filter is therefore to drop details for records where the observed number of significant results is less than might be expected, given the number of tests made and the probability threshold being used. Thus if there are m tests and the probability threshold is 0.01 then m*0.01 significant results would be expected under the null hypothesis. Unless the reported number of significant results exceed, or come close to, this value then there may be little value in reporting them. This rule of thumb is unaffected by lack of independence between the tests but there is no easy way of computing the associated variance and it cannot be developed further without considerable, perhaps unwarranted, computational effort.

Details of the Monte Carlo significance test procedure

The most general version of STAM requires no use of population at risk information. It is argued that this type of information is rarely available and, if it does exist, it is often out of date and thus incompatible with the other data in the GIS. The simplest test statistic is a count of records that lie within a tri-search region. The problem is that this measure is influenced by spatial variations in the underlying population at risk distribution. Even if there is no data for it, it will almost certainly not be spatially uniform; or even temporally uniform. Under these circumstances there are three possible approaches to detecting unusual pattern.

1. Randomly locate the central case anywhere in the study region and compute the probability of larger numbers of cases being found that satisfy the search criteria. This is a global test and will be study region dependent. Maybe it will work best for small, relatively homogeneous study regions.
2. Fix the spatial component and compute the probability of obtaining similar or more extreme counts of records based on random samples of k records from the database; k is the number of records that lie within the geographic search region. This assumes that purely spatial clusters are of no great interest and that it is either space-time or space-attribute or time-attribute or space-time-attribute clustering that is important.
3. A variation of option 2 would be to fix either the temporal or attribute dimension; or else compute the smallest probability out of all three.

It is argued that option 2 is probably the most relevant and flexible. Indeed, space still enters into the analysis because it localizes the search and interacts with the time and attribute data. On the other hand, this test will be less sensitive than one based on population at risk data and some might regard it as counterintuitive. It is argued that a loss of sensitivity is not a particular problem in terms of the proposed exploratory usage in a GIS context. It is in essence a generalization of the randomization approach that would be applied to Knox's space-time statistic if a Monte Carlo simulation approach were to be used. A further useful development is Besag's sequential Monte Carlo test (Besag, 1991). This reduces the number of simulations needed unless the test statistic is going to be significant or nearly so. This will save considerable amounts of CPU time as there is no point in knowing precisely how insignificant a particular result might be.

Using artificial life to search out patterns in data bases

Pattern seeking artificial creatures

The principal problem with the STAM is its computational clumsiness due to its crude grid search. It may miss patterns or detect too many, and usually takes far too long to be viable general purpose technology. Computer times are a function of the number of records available for analysis and this inter-acts in a highly nonlinear way with the Monte Carlo significance test. Clearly a more efficient search procedure is needed that can adapt to the data envi-ronment that is being searched without any need for arbitrary user-defined search parameters.

The logic underlying the artificial life (AL) literature is particularly appeal-ing; see Langton (1989). The AL view of AI emphasizes that intelligence is not necessarily achieved by imposing global top-down rules of logic and rea-soning (for instance, as provided by the STAM search), which then dictate the behaviour of what is now no longer an autonomous agent. Rather it is adaptive behaviour, defined as a much broader and flexible ability to cope with a complex, dynamic, unpredictable artificial world as a means of sur-vival. Beer (1990) writes: 'To me, this penchant for adaptive behaviour is the essence of intelligence: the ability of an autonomous agent to flexibly adjust its behavioural repertoire to the movement-to-movement contingencies which arise in its interaction with its environment' (p. xvi). Artificial life offers an opportunity to synthesize pattern and relationship seeking life forms that may be relevant to GIS analysis; see Openshaw (1992). Such life-forms have no biological relevancy but maybe they have much to offer as intelligent pattern spotters. The emphasis is on bottom-up rather than top-down model-ling, on local rather than global control, simple rather than complex specifi-cations, and emergent rather than prespecified behaviour.

The ideas borrowed from AL are very relevant to improving the STAM search process. What is needed is an intelligent life-form that can explore the GIS database looking for evidence of pattern in any or all of the three data spaces. It can be driven by a genetic optimization procedure which can breed new AL forms designed to feed on pattern and reproduce only if they are successful in the data environment in which they live. The most able forms will survive, the weaker and less successful ones will die out. The hope is expressed that these artificial life forms will evolve over a number of gen-erations in ways that are best able to find and highlight any major patterns that are sufficiently important to be of potential interest, if indeed there are any. Mapping and following, perhaps by computer movie, the life and times of these artificial life forms may well provide fascinating insights into pattern and process; combining animation with locally adaptive autonomous intelli-gent creatures as they explore the massively complex space-time-attribute world presented by a GIS database.

Once one starts to think along these lines, it is only a small step towards adopting AL analogies as the basis for a whole new generation of intelligent spatial analysis creatures with a view of spatial analysis being built-up from this almost atomic level. It should be feasible to create various types of arti-ficial life forms that can roam around databases at night (using spare CPU time), looking for pattern, 'foraging' for relationships as 'food', evolving as a result of interaction with their data environment, and by tracking their behaviour it may be possible to gain deep insights about structure and process, unobtainable by a more constrained or traditional approach. Indeed, in this way it may be possible to consider developing AL methods that seek to replicate the functions of many statistical operations but in a highly flexible way and with few assumptions that matter; see for example, Openshaw (1988). In short, AL thinking provides a potentially fruitful new paradigm for tackling many complex spatial analysis and modelling tasks.

An algorithm for creating a basic space-time-attribute creature (STAC/1)

The idea then is to create an artificial life form that searches for pattern. In essence a space-time-attribute creature (STAC) need be no more than a parameterized generalization of the STAM algorithm with the parameters themselves adapting to the database environment via a genetic algorithm. The resulting 'creature' is essentially a hypersphere that is of a flexible shape that can move in space and time. The STAC has the following parameters:

1. easting and northing coordinates defining its geographic location;
2. a temporal coordinate defining its position in time;
3. a vector of attributes defining its characteristics; and
4. dimensional information with maximum search radii for all three data spaces.

This hypersphere wanders around the database under the control of a genetic algorithm which seeks to minimize the Monte Carlo test probability that the number of data records captured by any specific AL–hypersphere is due to chance alone.

The basic algorithm is as follows.

Step 1. Define the genetic algorithm parameters, number of iterations, generation size and number of generations. Decide how to map the parameters that define the STAC onto a bit string for the genetic algorithm.

Step 2. Select initial starting locations.

Step 3. Evaluate their performance using a sequential Monte Carlo significance test.

Step 4. The best performers reproduce by applying genetic operators to pairs of bit strings to create a new generation of STACs.

Step 5. Repeat Steps 3 to 4 until the maximum number of generations are exceeded.

Step 6. Report the parameters of the best performing STAC and remove the associated data records from the database.

Step 7. Repeat steps 2 to 6 a set number of times.

The STAC is merely a generalization of the STAM but with an intelligent and adaptive search mechanism. The STACs hunt for minimum due-to-chance probability results. The form of the patterns they may find is represented by their configuration. The STAC offers the following advantages:

1. the metric of the search space is highly elastic and can be performed in an extremely precise manner; unlike the STAM in which a coarse computation saving grid had to be used;
2. it makes minimal assumptions about the nature of any pattern, the latter being self-evident from its configurational parameters;
3. the search process is intrinsically parallel, highly efficient and flexible enough to allow overlapping patterns;
4. it reduces the multiple testing problem by testing few hypotheses;
5. it uses significance only as a guide to the search;
6. data uncertainty effects could be included via the Monte Carlo testing procedure;
7. it is computationally highly efficient and suitable for workstation use on even very large problems;
8. it is very flexible and can be adapted to use whatever data exist; and
9. it assumes very little about the data or how it was generated.

Pattern in this context is defined as an unusually strong concentration of cases with similar but not necessarily identical characteristics. Suppose that the AL hypersphere is visualized as a multidimensional jellyfish, the idea being that its shape will engulf a region of the database, based around a particular location where something unusual is going on. Its radius in geographic space, its temporal dimension, and its attribute similarity envelope indicate where and what the unusualness consists of. The degree of unusualness can be defined by Monte Carlo simulation. Multiple AL forms can be created to roam around the database independently and those that prosper most will be grabbing locations where it may well be worth further investigation. This is similar to the STAM function that was discussed previously but the search spaces are now infinitely elastic, they need not be continuous, they could contain holes, they could be ring based, and they could be directional. Additionally, the creatures are searching many different locations in an intrinsically parallel manner. Because they are using local intelligence rather than a global search heuristic, they need far less computer time and they can, if they wish, indulge in a far more detailed search process than would ever be possible with a STAM/1 type of brute force heuristic.

Empirical Results

A crime pattern analysis problem

It is useful to demonstrate both the STAM and STAC on a GIS database. The example used here relates to the Crime Pattern Analysis task. It is assumed that crime data might contain useful insights about the existence of spatial and temporal pattern, if they can be identified quickly enough. This

Table 5.2. Crime database.

Variable	Number	Type
easting and northing	2	integer
day number*	2	integer
weekend flags*	2	0/1
modus operandi	12	0/1
crime type	11	0/1
day of week*	14	0/1
time of day*	6	0/1

Notes: *two values are needed; crimes are reported as occurring between two periods of time. The 0/1 recodes provide a common distance metric for the attribute data.

Table 5.3. STAM search parameters.

Run	Search space	Radii
1.	geographic	200, 400, 600
	temporal	2, 4, 6, 8, 10 days
	attribute	4, 8, 12 (mismatches out of 45)
	minimum size	3
2.	geographic	200, 400, 600, 800, 1000
	temporal	2, 4, 6, 8, 10 days
	attribute	2, 4, 6, 8, 12, 14, 16 (mismatches out of 45)
	minimum size	3
3.	geographic	200, 400, 600
	temporal	2, 4, 6, 8, 10 days
	attribute	4, 8, 12 (mismatches out of 45)
	minimum size	10

might be regarded as a generic real-time analysis problem involving data from the three principal data spaces likely to be found in a GIS. Indeed, the database used came from an experimental GIS operated by Northumbria Police in one of their Divisions for one year. Table 5.2 lists the data that are used. There is no data selection and all the available data are input although to simplify matters the attribute data was recoded as 0/1's. A crime pattern of interest would be an unusual concentration of events which are 'fairly' close in space, 'fairly' close in time, and 'fairly' similar in terms of their data characteristics. It is noted that the definition of 'fairly close' and the locations in both geography and time, plus the crime profile of the cases are left to the STAM and STAC procedures to define.

STAM/1 results

Three runs were performed. Table 5.3 shows the grid-search parameters that were used. Run 1 generated 28,845 searches, focused only on burglaries (n =

Table 5.4. Size distribution of STAM/1 results.

Number of significant results	Frequency
> 10	23
8–10	56
6–7	88
5	86
1–4	695
0	1,223

641) and required 5 hours CPU time on a SunSparc2. Run 2 examined the whole database (n = 2,171) used a more extensive search, generated 379,925 searches, and took 5 days to run. Run 3 used the same search parameters as Run 1 but processed the entire data set together with a minimum cluster size of 10 records in about 4 hours of CPU time. The reduced time reflects the smaller number of Monte Carlo simulations required because of the higher minimum size. Clearly this STAM method would greatly benefit from multiple-processors since it is a highly parallel search process. Also in practice, the computational load could be greatly reduced by only exploring either certain cases of particular interest or only part of the database (viz. the last two months), in which case the STAM would be being used mainly as a fuzzy spatial query tool.

Table 5.4 presents a summary of the results obtained from Run 3. It would seem that there are a relatively small number of locations around which the patterns are extremely strong such that the selection of search parameters was no longer critical. In Run 3 the maximum number of searches per case is 45, so at some locations 10 or more of the 45 permutations of search would yield probabilities less than the threshold of 0.01; indeed, many of the probabilities were close to zero. Table 5.5 reports details for one particular location. It should be noted that this technology cannot distinguish between useful and useless results and this responsibility is left to the analyst.

STAC results

The STAC is run on the entire database. A run to find the first ten patterns takes less than two hours on a workstation, with most of the time being for the first run. In fact the search usually identifies multiple optima and each run could be used to define several different locations at a time. The genetic algorithm parameters were set at: population size of 32, cross-over probability of 0.95, mutation probability of 0.01, and inversion probability of 0.05; see Goldberg (1988) and Holland (1975).

Table 5.6 summarizes the results obtained. The smallest probability values are limited by the maximum number of Monte Carlo simulations which is

Table 5.5. Typical STAM results for one search location.

east	north	times		data attributes
205	289	460	461	100010000000010000000000000000011000000001100
504	926	459	461	100010000000000000001000000010100000010010
486	083	459	461	1000100000001000000000000000010100000001100
118	124	468	468	000010000000010000000000010000001000000010001
228	194	470	471	000010000000000000000000010000000100000010100
**196	181	474	475	100010000000000000000000000000011000000010100
385	307	474	475	1000100000001000000000000000001100000001100
321	246	473	475	100010000000000000000000000010100000001010
244	228	468	475	000010000000000000010001000001000000010010
242	237	473	475	1000100000001000000000000000010100000010010
241	217	474	476	100010000000010000000000000010100000010100

Notes: search around case flagged ** for 400 metres, +/– 15 days and 6 attribute mismatches.
This particular location was also significant for the following combinations of search parameters:

geographic space	time in days	attribute mismatch
1,200 m	15	6
1,200 m	10	6
1,200 m	5	6
1,100 m	15	6
1,100 m	10	6
1,100 m	5	6
1,000 m	15	6
1,000 m	10	6
1,000 m	5	6
800 m	15	6
600 m	15	6
600 m	10	6
400 m	15	6

currently set at 1,000. The STAC is highly efficient in that it obtains the best possible results (i.e. $p = 0.001$) very quickly indeed; usually after about 100 evaluations. Clearly it is not difficult to find potential patterns although whether they make any sense is a matter for subsequent study. One problem in a crime pattern analysis context is that there no known examples of pattern being found due to the difficulties of the task. In other words there are no benchmarked results available. All that is known here is that the database was seeded with some fuzzy pattern and the STAC finds it relatively easily.

Some clues as to the nature of the patterns are also given in Table 5.6. The search radii suggest the nature of the patterns being identified. In a live situation it would probably be necessary to restrict the STACs to data that was recently acquired with less emphasis on the old data values.

Table 5.6. Typical STAC results.

Run number	Best p	Number of points	Easting	Northing	Time	t-r	d-r	a-r
1	0.001	45	727	788	423	16	990	8
2	0.001	8	584	407	227	16	1,020	5
3	0.001	9	588	476	432	11	1,230	6
4	0.001	7	566	988	235	18	5,360	3
5	0.006	147	594	461	197	20	2,150	9
6	0.001	6	646	527	409	25	1,450	3
7	0.001	19	764	511	453	16	1,810	6
8	0.001	7	072	182	251	10	2,490	6
9	0.001	40	592	465	403	9	5,350	6
10	0.001	7	604	339	427	16	1,370	6

Notes: t-r is time radius in days, d-r is geographic space radius in metres and a-r is the maximum attribute mismatch.

Some comments

The space-time-attribute cluster analysis being performed here is new territory for spatial analysis. It is made feasible by a fast computer, either brute force or smart search algorithms, and a computational philosophy which states that such things are worthwhile. The knowledge and insights that may be generated are currently unobtainable by any other route. Although the intelligence is limited to the repeated application of an extremely simple search and analysis procedure, it may nevertheless be fairly powerful because it can be applied to a very large number of potential pattern locations under the control of an exhaustive search heuristic. The intelligence added by the computer, results from being able to conduct this extensive search process, thereby answering the questions 'where' might there be some evidence of pattern, and 'what' characteristics do these patterns possess. By examining many different locations in a complex, multidimensional, and multivariable geographic-temporal-attribute space it may be possible to uncover new knowledge about pattern and process via mapping the results and applying human insight. The real end-user intelligence is still provided by the human-being but the evidence on which it is based is now computer generated.

Conclusions

It may seem that given the current state of neglect of spatial analysis in GIS that the methods being advocated here are far too sophisticated. However, it is argued that whilst there is some truth in this statement, the analysis task might actually be easier when viewed from this broader perspective. Additionally, it may be useful to have an image of what might be considered to be the ultimate in exploratory pattern analysis technology relevant to GIS, because only then is it likely that the relevant tools will be perfected. Indeed, quite often the hardest part of the problem is simply knowing what the problem is. Of course at present there will be no concensus that anything along the lines proposed here is necessary. This view is quite understandable but also lacking in both imagination and any appreciation of where the ongoing developments in AL and computer hardware speeds are likely to take us and how they may well impact on spatial analysis thinking in GIS environments. This chapter has described some of the ways by which first generation tri-space analysis tools can be built and it has presented the results of an empirical test. This is not meant to horrify but to stimulate debate about how best to analyse GIS databases.

Of course, it can be argued that the results are almost impossible to evaluate. This is not surprising given the dimensionality restrictions, even blindness, that characterizes much spatial analysis. Researchers have attempted to keep things simple and by being too simple minded they are open to the criticism that they may be missing too much by only ever analyzing part of the available information. It is also important to realise that pattern in tri-space may well look very different from that historically found in any of the spaces considered separately. Indeed, many examples of supposedly well-known patterns may in fact be generated by data selection decisions. For example, the famous Sellafield cancer cluster is tight in a geographic domain but spread over a very long time period and composed of disease types which are non-identical but generally lumped together. It may not be real, but generated as an unintended artefact of the analysis protocols. On the other hand, tri-space patterns may well be so complex that they are difficult to interpret because of our lack of experience in this area. At the same time it is essential to look, if only because there is increasing evidence that purely spatial patterns are hard to find and when they are found there is increasing uncertainty as to whether they have any diagnostic value; for example, in spatial epidemiology the expected major breakthroughs have not yet materialized and maybe this is because the analyses are too unidimensional. Indeed there is a feeling that there may well be many very interesting and insightful discoveries to be made once the appropriately powerful and restriction-free exploratory analysis techniques have been perfected.

The technology being advocated here is not meant to compete with more conventional statistical analysis methods. It is merely designed to operate in domains where standard methods cannot at present function and to act as a front-end to the analysis process. The aim is to explore GIS databases, to create new insights, and generate suggestions for more focused and detailed investigations. It is also argued that exploration cannot be a purely statistical process in a narrowly defined way. Rather it should be a more fluid, artistic, and creative activity. The analysis tools have to be able to stimulate the user's imagination and intuitive powers, as well as being able to handle the unexpected and to search everywhere. The need for such methods can be justified if they can be developed into a safe technology able to produce useful results in a reliable manner from the masses of gathered spatial information. It is very early days yet but there is some evidence of considerable promise.

Acknowledgement

Thanks are due to the ESRC for funding the North East Regional Research Laboratory where many of the developments briefly described were developed.

References

Beer, R. D., 1990, *Intelligence as Adaptive Behaviour: An experiment in Computational Neuroenthology*. Academic Press.

Boston. Besag, J., Newell, J., 1991, The detection of clusters in rare disease, *Journal of the Royal Statistical Society Ser. A*, **154**, 143–156.

Besag, J., Newell, J., Craft, A., 1991, The detection of small-area anomalies in the database, G. Draper (Ed.), The Geographical Epidemiology of childhood leukaemia and non-Hodgkin lymphomas in Great Britain, 1966–83; *Studies in Medical and Population Subjects No. 53*, OPCS, HMSO, London, pp. 101–108.

Diggle, P. J., 1991, A point process modelling approach to raised incidence of a rare phenomenon in the vicinity of a prespecified point, *Journal of the Royal Statistical Society A*, **153**, 349–362.

Goldberg, D. E., 1989, *Genetic Algorithms in Search, Optimisation, and Machine Learning*, Reading, Mass.: Addison-Wesley.

Holland, J. H., 1975, *Adaptation in Natural and Artificial Systems*, Ann Arbor: The University of Michigan Press.

Knox, E. G., 1964, The detection of space-time interactions, *Applied Statistics*, **13**, 25–29.

Langton, C. G., 1989, Artificial Life, C. G. Langton (Ed.), *Artificial Life*, Redwood, Calif.: Addison-Wesley, pp. 1–48.

Mantel, N., 1967, The detection of disease clustering and a generalised regression approach, *Cancer Research*, **27**, 209–220.

Marshall, R. J., 1991, A review of methods for the statistical analysis of spatial patterns of disease, *Journal of the Royal Statistical Society Ser. A*, **154**, 421–442.

Openshaw, S., 1988, Building an automated modelling system to explore the universe of spatial interaction models, *Geographical Analysis*, **20**, 31–46.

Openshaw, S., 1989, Computer modelling in human geography, B. Macmillan (Ed.), *Remodelling Geography*, Oxford: Blackwell, pp. 70–89.

Openshaw, S., 1990a, Spatial analysis and GIS: a review of progress and possibilities, H. Scholten and J. C. H. Stillwell (Eds.), *Geographic Information Systems for Urban and Regional Planning*, Dordrecht: Kluwer, pp. 153–163.

Openshaw, S., 1990b, Automating the search for cancer clusters: a review of problems, progress, and opportunities, in R. W. Thomas (Ed.), *Spatial Epidemiology*, London: Pion, pp. 48–78.

Openshaw, S., 1991, Developing appropriate spatial analysis methods for GIS, D. Maguire, M. F. Goodchild, D. W. Rhind (Eds.) *Geographical Information Systems: principles and applications*, London: Longman, pp. 389–402.

Openshaw, S. and Craft, A., 1991, Using geographical analysis machines to search for evidence of clusters and clustering in childhood leukaemia and non-Hodgkin lymphomas in Britain, G. Draper (Ed.), The Geographical Epidemiology of childhood leukaemia and non-Hodgkin lymphomas in Great Britain, 1966–83; *Studies in Medical and Population* Subjects No. 53, OPCS, HMSO, London, pp. 109–122.

Openshaw, S., 1992, Some suggestions concerning the development of AI tools for spatial analysis and modelling in GIS, *Annals of Regional Science*, **26**, 35–51.

Openshaw, S., Charlton, M., Wymer, C. and Craft, A., 1987, A mark 1 Geographical Analysis Machine for the automated analysis of point data sets, *International Journal of GIS*, **1**, 335–358.

Openshaw, S., Cross, A. and Charlton, M., 1990, Building a prototype geographical correlates exploration machine, *Int. Journal of GIS*, **3**, 297–312.

Openshaw, S., Wymer, C. and Cross, A., 1991, Using neural nets to solve some hard problems in GIS, *Proceedings of The Second European Conference on Geographical Information Systems*, EGIS Foundation, Utrecht, The Netherlands, pp. 797–807.

6

Spatial dependence and heterogeneity and proximal databases

Arthur Getis

The solution to spatial problems in a geographic information systems environment inevitably must be based on the special character of spatial data (Anselin and Getis, 1991). In recent years there has been a spate of expository tracts that have detailed the attention that must be given spatial data if they are to be used in analysis. In the last several years such authors as Anselin (1988), Cressie (1991), Upton and Fingleton (1985), Haining (1990), Odland (1989), Goodchild (1987) and Griffith (1988) have clarified the statistical and econometric issues involved in the use of spatial data. This area may be called Spatial Data Analysis (SDA). At the same time, the Geographic Information Systems (GIS) field has witnessed a plethora of writing on the use of spatial data in a computer environment. Recent work by Burrough (1990), Goodchild and Kemp (1990), and Samet (1987) has gone far to organize the literature in this field. Thus, there are two areas, statistical and computer-based, that owe their existence to the nuances of spatial data. The main goal of the first is to model and analyse spatial data, the main goal of the second is to organize and manage spatial data. The fact that the overlap between the two is minor is a function of these main goals.

Several commentators on SDA have attempted to identify the entry points for SDA into GIS environments (Anselin and Getis, 1992; Haining and Wise, 1991; Openshaw, 1991). In fact, several well-known GIS provide functions, modules, or macro-languages that allow for modest analytical capabilities (ARC/INFO, IDRISI, SPANS). In SDA, the heteroskedasticity and dependence (interdependence) of spatial data receive most attention.

Broadly speaking, *heterogeneity* in a spatial context means that the parameters describing the data vary from place to place. For an arbitrary stochastic process the distribution function depends on distance between elements with the parameters characterizing it being functions of distance. Non-stationarity, or spatial heterogeneity, occurs when the process observed in a window (or kernel) changes systematically. This results from either the presence of a trend in the data or a change in the variance.

If the distribution function of the stochastic process remains unchanged when distance changes by an arbitrary amount then the process is stationary and *spatially homogeneous*. That is, space-homogeneity is restricted to be a function of the distance between the elements of the distribution in question. *Spatial dependence* is a special case of spatial homogeneity. It implies that the data for particular spatial units are related *and similar* to data for other nearby spatial units in a spatially identifiable way.

GIS and spatial data

While the computer-based or GIS emphasis is on technical questions of spatial data handling for various system architectures and data structures, a recent summing-up by Sinton (1992), who has worked in GIS for 25 years, identifies spatial dependency as the chief area for GIS research in the future.

> Now we have discovered that the essential nature of geographic data creates both a fundamental theoretical problem and a nasty technical problem that uniquely defines the industry . . . I believe the spatial inter-dependence among geographic entities is the theoretical linchpin of the GIS industry. [The technical problem Sinton identifies is the apparent limitations imposed by irregular and variable data; see Sinton (1992), pp. 2–3]

It follows that there is at least one area of SDA, spatial dependence, that is of central concern to those working in GIS.

In this chapter, I would like to outline a procedure whose outcome would be an enhanced capability for GIS to display and aid in the evaluation of spatial dependence and heterogeneity. I suggest that data that are attributed to a particular point, line, or area be used to derive information on spatial qualities, and that the derived data be considered as attributes of the spatial unit. Our interest is in extending data bases to include statistical information about spatial units in the vicinity of the spatial unit in question. The purpose is to emphasize an extended view of space to include both 'site' and 'situational' characteristics of spatial units, and to alert researchers to the spatial nuances of their data. This point of view has been expressed eloquently by Couclelis (1991) who recognizes that the usefulness of a GIS, especially for the urban planning fields, is dependent on an image of space as relational, somehow expanding and deepening our conception of space to something intermediate between site and situation, called *proximal space*. As important as it is to know the site attributes of data units, so, too, is it important to know of the nature of a spatial unit relative to other spatial units. Especially, as new technology allows for an increasing number of spatial units representing smaller and smaller portions of the earth's surface, questions of data compression, data selection, and the quality of analytical work with such data become important. This point of view is directly related to the question of the appro-priateness of using a particular spatial scale in analytical work.

The importance of knowledge about spatial dependence cannot be over-emphasized. The moments of distributions, spatial or non-spatial, are the key to making inferences about what data represent. In a mosaic of spatial units, the variance of an attribute leads to an understanding of its dependence structure. Without that knowledge, any opinion about the meaning of the spatial unit attribute is suspect. Armed with knowledge of the nature of spatial dependence, appropriate models may be specified about the relationships embodied in the data.

The proximal theme

By emphasizing the importance of knowing the dependence structure of a spatial data set, I am suggesting that included within a typical data base is information on spatial dependence. The determination of any measure of spatial dependence implies that the attributes of more than one spatial unit are required. This implies that the characteristics of units relative to the characteristics of other spatial units in geographical proximity must be stored in some convenient way. The theme is proximal in the sense that attribute data contain information not only on the georeferenced site but on the relationship of the site attributes to the attributes of georeferenced data in the vicinity of the site. The question arises, of course, as to which units are to be included in a measure of dependence. I suggest a flexible proximal system. The idea is related to the kernel approach of density estimation (Diggle, 1985; Silverman, 1986) and various smoothing algorithms (Foster and Gorr, 1986).

Spatial dependence and heterogeneity descriptors

Those concerned with spatial dependence have a number of tools at their disposal that help describe or model variance structure. The typical measures include parameters of variables or error terms in spatially autoregressive systems, Moran's I, Geary's c, semi-variogram parameters, or spatial adaptive filter parameters. In addition, various econometric tests for heteroskedasticity, such as the Breusch-Pagan (1979) test, can be adapted for spatial situations (Anselin, 1988). Although not straightforward, all of these measures can be entered into a GIS via kernel routines. The question arises as to the storage location of this information. What is proposed here is that spatial dependence and heterogeneity descriptors be stored as attributes of individual data units.

Variance

In the work that follows, values of the variance greater than that which would be expected in a sample drawn from a population with mean μ and variance σ^2 are identified as representative of heterogeneous circumstances. Statistically significant consistency in the variance (little variance), that is, less variance than that which would be expected in a set of data which has been sampled from a population having mean μ and variance σ^2 represents a spatially dependent circumstance. A technique is introduced in the next section that, combined with the study of variance, allows for a finer classification of the data – one that allows us to make statements about spatial trends in the data, that is, trend heterogeneity.

From the above, we define four types of spatial patterns: spatially dependent observations, heterogeneity patterns due to extreme variance, heterogeneity patterns due to spatial trends (both upward and downward), and patterns that result from the expected random fluctuation in the stochastic process (stationarity) that represents the body of observations that make up the population.

Trend

Following from the work of Ripley (1981), Getis and Franklin (1987) devised a test for scale effects using a neighbourhood approach. Each individual data unit is the node for measurements of density for increasing areas centred on the node. This work stimulated Getis and Ord (1992) to develop a statistic G_i^* based on the expected association between weighted points within a distance d of a place i. The statistics' value then becomes a measure of spatial clustering (or non-clustering) for all j within d of i. When the statistic is calculated for each d and for all i, one has a description of the clustering characteristics or SD of the area in question. It must be clear that the information at i is included in the measure:

$$G_i^*(d) = \frac{\sum_{j=i}^{n} w_{ij}(d)x_j}{\sum_{j=1}^{n} x_j}$$

where $\{w_{ij}\}$ is a symmetric one/zero spatial weight matrix with ones for all links defined as being within distance d of a data unit i; all other links are zero. The numerator is the sum of all x_j within d of i. This is a measure of the SD within d of i. If values of G_i^* are lower or higher than expected in a random distribution at distance d, SD obtains. The degree to which G_i^* varies from the expected is an expression of the degree of spatial dependence in the data. If one notes the changes in G_i^* as d increases, trends in the data can be identified. Details on tests and the moments of the distribution of the statistic can be found in Getis and Ord (1992).

What is suggested here is that the information obtained from the statistic $G_i^*(d)$ be stored as an attribute of each spatial unit. The data may be stored in any of a number of ways: (1) as one measure for a particular d, (2) as a series of measures for a number of d values, or (3) as a function of d. With regard to (3), the parameters of an equation that describe the $G_i^*(d)$ for many values of d are measures of trend for each i. The parameters can be those for

Figure 6.1.

a best fitting curve within the kernel for some relatively simple function (linear, quadratic, logistic, etc).

Regions having similar parameters may be delimited and stored efficiently. When a raster data structure is employed, a quadtree storage system of attribute data would be based on the parameters stored for each cell (Samet, 1990). Similar parameter values among a group of contiguous cells would be stored, in general, at one or a few locations in the spatial data base management system. Such an approach is not dependent on a data structure but would be influenced by it.

Examples using pixel data

Our two examples are based on a remotely-sensed image of a rapidly growing coastal area north of the city of San Diego in Oceanside, California. The area is pockmarked with highly reflective areas – shopping centres, industrial parks, areas that have been bulldozed, waiting for the construction of houses (Figure 6.1). Most of the remaining region is covered in either coastal sage

53	52	51	50	49	43	34	33	31	29	31	36	43	45	47	47
53	54	52	50	47	40	32	32	32	33	33	37	46	48	49	50
55	56	54	53	48	39	32	33	37	41	44	47	52	51	51	51
56	55	54	52	49	41	33	32	38	45	51	53	53	52	52	52
56	56	56	53	49	44	35	30	33	40	50	51	52	52	52	53
56	56	55	53	50	46	37	32	36	43	46	48	51	53	54	52
58	58	54	53	52	47	40	34	36	41	41	46	50	49	49	51
55	56	54	54	53	48	41	34	34	34	36	44	39	43	49	52
54	54	55	54	52	49	44	37	33	32	34	40	40	44	50	52
54	55	56	53	52	50	47	41	36	36	43	48	47	49	49	50
51	51	53	53	52	51	49	42	35	42	45	48	49	48	46	43
50	51	52	53	51	51	50	46	39	39	40	39	40	42	43	47
42	43	46	49	51	50	50	49	41	33	34	34	35	41	46	49
43	35	37	41	45	46	47	47	40	32	31	31	39	48	50	47
38	35	37	39	40	39	40	41	35	31	32	38	48	51	51	45
39	36	32	36	37	35	33	35	34	34	39	46	53	47	43	43

Figure 6.2(a).

scrub or residential housing. There are, however, a number of deep canyons where vegetation is dense with some trees among the sage.

Region I: a relatively homogeneous area

The first example is taken from the extreme northwest part of Figure 6.1. It is an area of reasonably homogeneous coastal sage scrub vegetation. We have selected a rectangle of 8 by 8 pixels from the centre of a 16 by 16 segment of the 512 by 512 pixel image where each pixel covers an area of approximately 30 metres by 30 metres. Each pixel has a light value on the 0 to 255 grey scale (Figure 6.2(a)). The region appears to have two somewhat different densities of vegetation, one with lighter values in the 50 to 60 range (sparse brush) and the other in the darker 30 to 50 range (thick brush) (see Figure 6.2(b)). Use of the central 64 pixels obviates the need to address the effect of boundaries. For these examples we will assume that the distance separating pixels is one, both horizontally and vertically. Thus, the distance to the nearest boundary of the 16 by 16 region from the top left pixel of the study area is 4 units. We select a window of distance 4 around each ith pixel.

To identify significant departures from the expected variance, a large sample of pixel values taken from the entire area in Figure 6.1 was collected. The

Figure 6.2(b).

variance for these, 238.65, represents the population variance. For each pixel in Region I, all pixel values within distance 4 of each pixel were used to calculate the sample variances. We then employ a one sample analysis of variance. The sample variances should be distributed as chi-square (with the appropriate degrees of freedom) if the samples were taken from a normal population. Since all samples in this study have 48 degrees of freedom, one would not expect sample values greater than 65.15 at the 95 per cent level and less than 33.11 at the 5 per cent level. These values are depicted on the Y axis in Figure 6.3. Also given in Figure 6.3 are the chi-square values for the 99.5 per cent and 0.5 per cent levels. Values above the 95 per cent level are considered to represent heterogeneous samples while values below the 5 per cent level represent samples that are spatially dependent. Figure 6.4 shows that all of the values for Region I represent spatial dependence.

The G_i^* value for each i when d equals 1, 2, 3, and 4 is given in Figure 6.5. The expectations are the G_i^* values for each d that would obtain if the values of all pixels within distance d are exactly the same as the sample pixel i. The G_i^* values for each of the 64 pixels for each of four distances become the

A. Getis

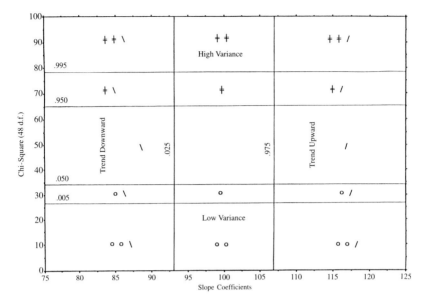

Figure 6.3.

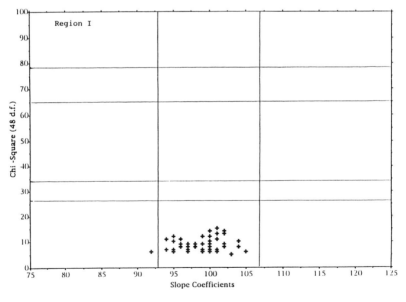

Figure 6.4.

d=1

0.0214	0.0188	0.0156	0.0142	0.0155	0.0184	0.0208	0.0222
0.0219	0.0196	0.0166	0.0148	0.0157	0.0180	0.0199	0.0212
0.0223	0.0204	0.0174	0.0154	0.0158	0.0170	0.0184	0.0200
0.0226	0.0208	0.0181	0.0157	0.0149	0.0155	0.0165	0.0179
0.0227	0.0212	0.0191	0.0165	0.0150	0.0148	0.0162	0.0180
0.0226	0.0218	0.0202	0.0177	0.0158	0.0165	0.0180	0.0198
0.0226	0.0221	0.0209	0.0186	0.0170	0.0172	0.0191	0.0200
0.0226	0.0221	0.0215	0.0198	0.0175	0.0169	0.0172	0.0176

d=2

0.0549	0.0483	0.0423	0.0399	0.0426	0.0476	0.0522	0.0559
0.0560	0.0502	0.0441	0.0409	0.0420	0.0459	0.0507	0.0549
0.0569	0.0519	0.0460	0.0419	0.0412	0.0436	0.0480	0.0509
0.0576	0.0533	0.0476	0.0427	0.0406	0.0418	0.0446	0.0475
0.0584	0.0546	0.0494	0.0441	0.0407	0.0418	0.0442	0.0473
0.0588	0.0559	0.0515	0.0463	0.0434	0.0435	0.0461	0.0491
0.0586	0.0566	0.0531	0.0493	0.0458	0.0446	0.0465	0.0492
0.0577	0.0568	0.0544	0.0506	0.0465	0.0438	0.0443	0.0458

d=3

0.1193	0.1087	0.1004	0.0970	0.0980	0.1033	0.1115	0.1196
0.1216	0.1111	0.1028	0.0979	0.0975	0.1022	0.1096	0.1171
0.1242	0.1138	0.1049	0.0984	0.0961	0.0993	0.1055	0.1123
0.1259	0.1168	0.1078	0.1004	0.0968	0.0985	0.1037	0.1100
0.1274	0.1195	0.1103	0.1029	0.0989	0.0995	0.1028	0.1082
0.1290	0.1220	0.1133	0.1065	0.1017	0.1001	0.1011	0.1054
0.1280	0.1231	0.1163	0.1093	0.1040	0.1005	0.0996	0.1030
0.1245	0.1219	0.1166	0.1100	0.1042	0.1000	0.0988	0.1025

d=4

0.1995	0.1870	0.1778	0.1712	0.1685	0.1721	0.1803	0.1916
0.2025	0.1898	0.1797	0.1732	0.1710	0.1735	0.1808	0.1914
0.2059	0.1928	0.1816	0.1745	0.1724	0.1745	0.1806	0.1907
0.2089	0.1960	0.1842	0.1769	0.1734	0.1741	0.1797	0.1886
0.2109	0.1990	0.1882	0.1805	0.1759	0.1741	0.1773	0.1836
0.2116	0.2021	0.1923	0.1835	0.1775	0.1736	0.1732	0.1773
0.2085	0.2016	0.1927	0.1840	0.1770	0.1723	0.1704	0.1736
0.2030	0.1974	0.1892	0.1814	0.1748	0.1711	0.1713	0.1758

Figure 6.5.

input data for the measure of spatial trend. Since we are dealing with an increasing radius as distance increases, we fit the four values by a square root function. If all values in the vicinity of a given pixel are balanced about the mean, the slope of the least squares lines is 0.10045 and the standard error is .00351. As expected, the fit is excellent (r = .9988). The results for the slope coefficients are given in Figure 6.6.

Using the 95 per cent confidence interval, the slope significance values are 'greater than 0.10733' and 'less than 0.09357'. Values greater or less than these represent upward or downward trending values, respectively. The areas of significant slope are identified on Figure 6.3. The symbols shown on Figure 6.3 represent the various sectors of the diagram. These symbols are used in Figures 6.4 and 6.7 to represent the proximal qualities of each pixel of Region I. The greatest deviation from the expected slope value is only 8 per cent.

A. Getis

Slope Coefficients

Expectation= 0.10045

```
0.10516  0.10606  0.11311  0.11998  0.12274  0.11978  0.11646  0.11596
0.10595  0.10633  0.11227  0.11982  0.12313  0.11748  0.11281  0.11396
0.10435  0.10274  0.10708  0.11487  0.11756  0.11067  0.10618  0.10976
0.09916  0.09822  0.10106  0.10770  0.11039  0.10431  0.10052  0.10617
0.09572  0.09564  0.09752  0.10166  0.10316  0.10002  0.09831  0.10324
0.09552  0.09931  0.09932  0.09920  0.09848  0.09779  0.09702  0.09951
0.09792  0.10072  0.09962  0.09750  0.09663  0.09771  0.09777  0.09627
0.09429  0.09488  0.09362  0.09199  0.09233  0.09462  0.09631  0.09427
```

Figure 6.6.

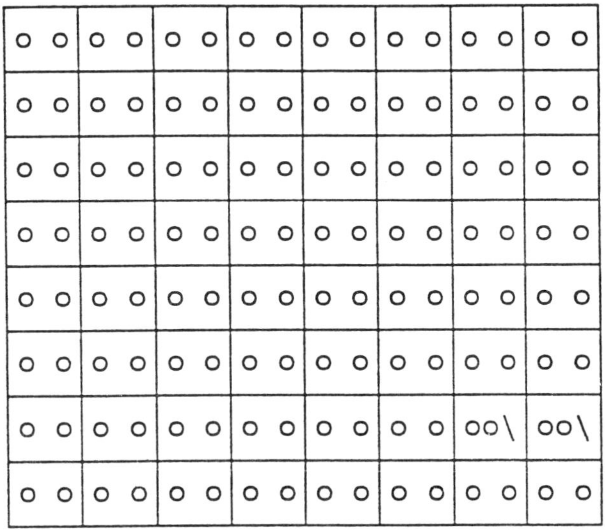

Legend:

$\ddagger \ \ddagger$ Variance > .995 → Great Heterogeneity
$\ddagger$ Variance > .95   → Heterogeneity

o   Variance < .05   → Dependence
o o   Variance < .005 → Considerable Dependence

\   Slope   < .025 → Trend Downward
/   Slope   > .975 → Trend Upward

Figure 6.7.

The implication of this is that there is relatively little variation in the pattern. All but two of the pixels are exclusively spatially dependent. The two pixels that are located at the beginning of a downward trend may be thought of as simultaneously spatially dependent and heterogeneous. No pixels are representative of the region as a whole. This information can now be easily stored (perhaps using a quadtree approach). Future sampling for confirmatory types of analyses (see Anselin and Getis, 1992) can be carried out with the knowledge that at the scale chosen (d = 4) the data are significantly spatially dependent.

Region II: an area of great disparity

Figures 6.1 and 6.8(a) identify the location and pixel characteristics of the second study region (Region II). Clearly, this area is much more diversified than the first region. Pixel values vary from very light (over 100) to moderately dark (in the 40s). Figures 6.8(a) and 6.8(b) show this contrast graphically. Figure 6.9 shows the G_i values for the 64 pixels from d = 1 to 4. Figure 6.10 shows that the largest slope departures are 23 per cent greater than the expectation. This value is nearly three times larger than in the first study region. Clearly, there is considerably more variation in the proximal data in this region than in the first study region. Figure 6.11 graphically portrays the heterogeneity. Finally, Figure 6.12 gives the final classification based on the scheme in Figure 6.3.

Three types of proximal data patterns predominate: the top portion of the study area is heterogeneous mainly due to an upward trend in the data, the bottom part is heterogeneous due to high variance, the remaining portion is representative of stationary data.

Conclusion

In summary, we are suggesting that one or more measures of spatial heterogeneity and dependence become a part of data sets where appropriate. Knowledge of these will:

- allow for an assessment of scale effects;
- help identify appropriate specifications for explanatory models;
- aid in the development of reasonable sampling schemes;
- facilitate data storage and compression; and
- help to delimit areas of similar proximal data structure.

The G_i^* statistic together with an analysis of variance measures spatial association in the neighbourhood of the spatial unit in question. By taking many measures, one for each d distance, a function for a window can be derived which describes how the statistic varies with distance from the spatial unit. The function can be considered as an attribute of the spatial unit. The

55	56	54	58	65	75	82	77	74	74	69	61	62	71	73	63
62	63	64	59	85	88	95	106	110	99	89	82	79	84	97	79
55	55	56	60	91	95	86	98	115	105	110	107	101	89	85	68
55	54	53	61	82	102	88	93	96	94	110	114	109	103	92	68
59	58	60	64	88	99	82	81	71	80	89	89	89	102	104	75
63	57	58	58	77	92	82	71	59	90	105	92	79	98	110	83
62	55	56	56	80	90	99	88	64	91	112	94	76	91	100	85
62	59	55	61	99	97	93	80	65	87	107	80	59	60	67	66
65	62	68	72	102	94	90	83	74	81	96	69	52	50	51	54
62	62	86	85	55	59	64	72	75	70	70	62	66	61	55	57
52	59	61	56	41	40	43	44	46	48	50	52	68	69	60	61
42	43	44	43	42	41	42	44	43	43	44	47	58	59	55	61
44	41	39	42	44	43	42	42	42	43	43	49	56	53	53	61
43	42	40	42	42	42	41	42	42	43	42	53	66	61	51	62
40	42	41	40	43	49	46	42	42	43	43	49	59	62	53	62
40	41	42	43	49	54	47	44	42	44	43	46	52	56	56	61

Figure 6.8(a).

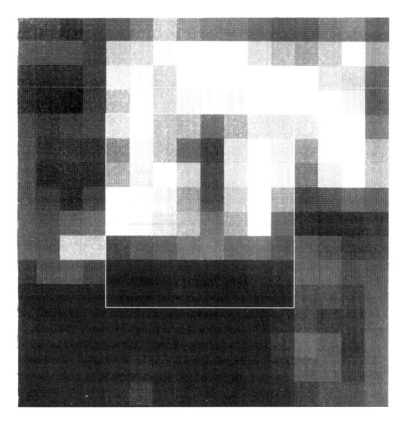

Figure 6.8(b).

d=1

0.0240	0.0272	0.0253	0.0233	0.0227	0.0249	0.0277	0.0277
0.0232	0.0258	0.0250	0.0223	0.0208	0.0249	0.0286	0.0269
0.0236	0.0269	0.0265	0.0236	0.0215	0.0260	0.0298	0.0266
0.0257	0.0277	0.0269	0.0240	0.0217	0.0253	0.0283	0.0240
0.0247	0.0259	0.0249	0.0234	0.0222	0.0239	0.0248	0.0211
0.0201	0.0183	0.0192	0.0198	0.0198	0.0202	0.0204	0.0187
0.0137	0.0131	0.0137	0.0146	0.0150	0.0151	0.0155	0.0164
0.0124	0.0122	0.0124	0.0126	0.0128	0.0130	0.0133	0.0147

d=2

0.0608	0.0658	0.0665	0.0638	0.0636	0.0680	0.0730	0.0738
0.0601	0.0655	0.0646	0.0617	0.0612	0.0648	0.0702	0.0708
0.0619	0.0672	0.0650	0.0613	0.0616	0.0659	0.0694	0.0671
0.0605	0.0666	0.0659	0.0623	0.0614	0.0643	0.0654	0.0613
0.0588	0.0609	0.0616	0.0588	0.0576	0.0596	0.0592	0.0550
0.0525	0.0518	0.0501	0.0495	0.0490	0.0513	0.0522	0.0494
0.0418	0.0389	0.0392	0.0402	0.0407	0.0416	0.0433	0.0441
0.0337	0.0332	0.0335	0.0344	0.0351	0.0355	0.0371	0.0401

d=3

0.1312	0.1395	0.1467	0.1510	0.1538	0.1569	0.1595	0.1616
0.1321	0.1389	0.1432	0.1464	0.1495	0.1495	0.1511	0.1551
0.1321	0.1383	0.1407	0.1431	0.1451	0.1429	0.1413	0.1441
0.1288	0.1324	0.1335	0.1373	0.1385	0.1359	0.1322	0.1332
0.1212	0.1242	0.1240	0.1261	0.1265	0.1252	0.1229	0.1223
0.1109	0.1135	0.1128	0.1098	0.1103	0.1112	0.1114	0.1103
0.0986	0.0980	0.0965	0.0951	0.0958	0.0981	0.1002	0.0996
0.0851	0.0835	0.0818	0.0825	0.0835	0.0865	0.0917	0.0934

d=4

0.2180	0.2299	0.2444	0.2573	0.2633	0.2621	0.2617	0.2596
0.2172	0.2267	0.2410	0.2540	0.2583	0.2554	0.2539	0.2513
0.2146	0.2199	0.2301	0.2430	0.2443	0.2398	0.2392	0.2389
0.2054	0.2103	0.2159	0.2243	0.2245	0.2194	0.2184	0.2206
0.1945	0.1976	0.2009	0.2064	0.2064	0.2038	0.2020	0.2034
0.1811	0.1848	0.1869	0.1893	0.1867	0.1870	0.1859	0.1865
0.1656	0.1694	0.1694	0.1680	0.1670	0.1699	0.1716	0.1712
0.1518	0.1530	0.1514	0.1482	0.1498	0.1554	0.1597	0.1603

Figure 6.9.

Slope Coefficients

Expectation= 0.10045

0.10516	0.10606	0.11311	0.11998	0.12274	0.11978	0.11646	0.11596
0.10595	0.10633	0.11227	0.11982	0.12313	0.11748	0.11281	0.11396
0.10435	0.10274	0.10708	0.11487	0.11756	0.11067	0.10618	0.10976
0.09916	0.09822	0.10106	0.10770	0.11039	0.10431	0.10052	0.10617
0.09572	0.09564	0.09752	0.10166	0.10316	0.10002	0.09831	0.10324
0.09552	0.09931	0.09932	0.09920	0.09848	0.09779	0.09702	0.09951
0.09792	0.10072	0.09962	0.09750	0.09663	0.09771	0.09777	0.09627
0.09429	0.09488	0.09362	0.09199	0.09233	0.09462	0.09631	0.09427

Figure 6.10.

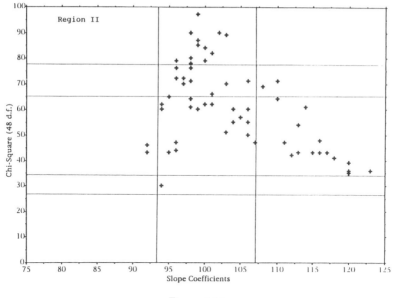

Figure 6.11.

way in which the statistic and the sample variance behave from unit to unit allows researchers the opportunity to assess the nature of spatial heterogeneity or dependence that may exist. Any researcher who uses a sampling scheme that selects pixels not classified as stationary must recognize that spatial dependence and heterogeneity may very well invalidate any further parametric analysis.

Acknowledgement

This work was done with the support of the National Science Foundation, Grant No. SES-9123832. The author appreciates the review of an earlier draft of this chapter by Gerard Rushton, and comments by Serge Rey and Cynthia Brewer.

References

Anselin, L., 1988, *Spatial Econometrics. Methods and Models*, Dordrecht: Kluwer.
Anselin, L. and Getis, A., 1992, Spatial statistical analysis and geographic information systems, *Annals of Regional Science*, **26**, 19–33.
Breusch, T. and Pagan, A., 1979, A simple test for heteroscedasticity and random coefficient variation, *Econometrica*, **47**, 203–7.
Burrough, P. A., 1990, Methods of spatial analysis in GIS, *International Journal of Geographical Information Systems*, **4**(3), 221.

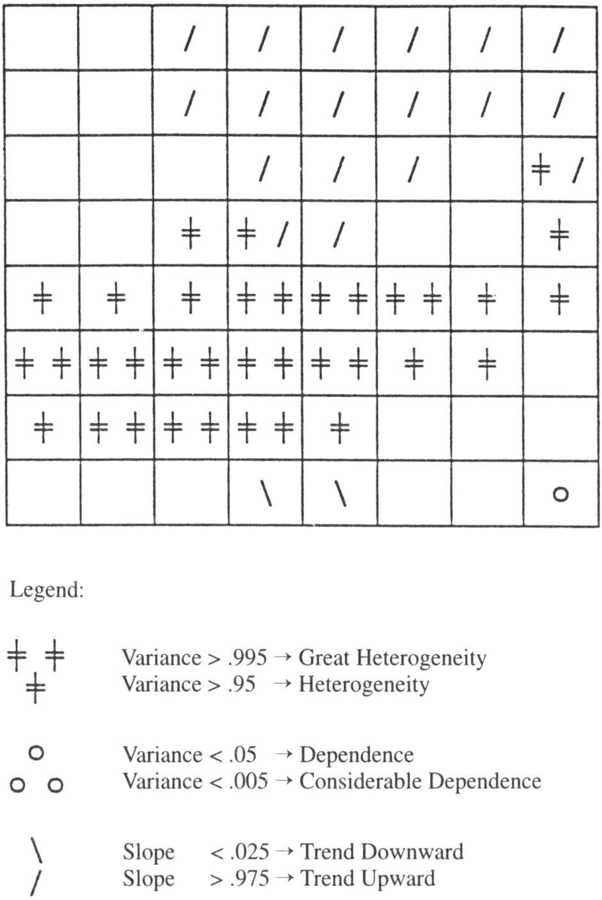

Legend:

‡ ‡   Variance > .995 → Great Heterogeneity
 ‡   Variance > .95 → Heterogeneity

 o   Variance < .05 → Dependence
o o Variance < .005 → Considerable Dependence

 \   Slope   < .025 → Trend Downward
 /   Slope   > .975 → Trend Upward

Figure 6.12.

Couclelis, H., 1991, Requirements for planning-relevant GIS: a spatial perspective, *Papers in Regional Science*, **70**(1), 9–19.

Cressie, N., 1991, *Statistics for Spatial Data*, Chichester: John Wiley.

Diggle, P. J., 1985, A kernel method for smoothing point process data, *Journal of the Royal Statistical Society (C)*, **34**, 138–47.

Foster, S. A. and Gorr, W. L., 1986, An adaptive filter for estimating spatially-varying parameters: application to modeling police hours in response to calls for service, *Management Science*, **32**, 878–889.

Getis, A. and Franklin, J., 1987, Second-order neighborhood analysis of mapped point patterns, *Ecology*, **68**, 473–477.

Getis, A. and Ord, J. K., 1992, The analysis of spatial association by distance statistics, *Geographical Analysis*, **24**(3), 189–206.

Goodchild, M. F., 1987, A spatial analytical perspective on geographical information systems, *International Journal of Geographical Information Systems*, **1**(4), 335–54.

Goodchild, M. F. and Kemp, K. K. (Eds.), 1990, NCGIA Core Curriculum, Santa Barbara, University of California.

Griffith, D., 1988, *Advanced Spatial Statistics: Special Topics in the Exploration of Quantitative Spatial Data Series*, Dordrecht: Kluwer.

Haining, R., 1990, *Spatial Data Analysis in the Social and Environmental Sciences*, Cambridge: Cambridge University Press.

Haining, R. and Wise, S. M. (Eds.), 1991, GIS and spatial data analysis: report on the Sheffield workshop, Regional Research Laboratory Initiative, Discussion Paper 11, University of Sheffield.

Odland, J., 1989, *Spatial Autocorrelation*, Newbury Park: Sage.

Openshaw, S., 1991, A spatial analysis research agenda, *Handling Geographical Information*, I. Masser, and M. Blakemore (Eds.), London: Longman.

Ripley, B. D., 1981, *Spatial Statistics*, Chichester: John Wiley.

Samet, H., 1990, *The Design and Analysis of Spatial Data Structures*, Reading, Mass.: Addison Wesley.

Silverman, B. W., 1986, *Density Estimation for Statistics and Data Analysis*, London: Chapman and Hall.

Sinton, D. F., 1992, Reflections on 25 years of GIS, insert in *GIS World*, February, 1–8.

Upton, G. J. and Fingleton, B., 1985, *Spatial Statistics by Example, Vol. 1: Point Pattern and Quantitative Data*, Chichester: John Wiley.

7

Areal interpolation and types of data

Robin Flowerdew and Mick Green

Introduction

A common problem in geographical or other regional research is the fact that the areal units for which data are available are not necessarily the ones which the analyst wants to study. This is a problem first of all for people interested in one particular type of unit, such as an administrative district, an electoral division or a sales territory. It becomes more of a problem for people wishing to compare data over time when boundaries of data collection zones are subject to change. For example, if electoral reapportionment has taken place between elections, it is difficult to compare the performance of candidates between the two elections; in general, the monitoring of developments of any kind becomes problematic if boundaries have changed. Even national censuses may be hard to compare over time because data collection units are subject to change (Norris and Mounsey, 1983).

Situations frequently occur where a researcher wants to compare a variable which is available for one set of zones with another variable only obtainable for a different, incompatible set. Common examples in the British context include comparisons between wards and parishes and between districts, constituencies and travel-to-work areas (reviewed by Walford *et al.*, 1989). It is often important to compare data collected by private or commercial organizations with data made available officially. For example, client or customer addresses or survey results may be compared with officially provided demographic or socioeconomic data. This requires comparisons between zones such as postcode areas or customer service areas with officially designated zones. A further commonly encountered problem is the comparison of data for officially constituted areas with data for distance bands around a city, store or pollution source. Sometimes the comparison may involve environmentally defined regions, such as geological or vegetation zones, stream catchments or elevation zones, whose boundaries will seldom match with those for which official data are available.

The comparisons mentioned above may be necessary for evaluating change, for assessing the performance of facilities or stores or for producing effective maps to show two or more phenomena in relation to each other. The most important type of use, however, is statistical testing for relationships between the variables concerned.

This problem is a fundamental one in geographical information systems where it is sometimes referred to as the polygon overlay problem. A preferable

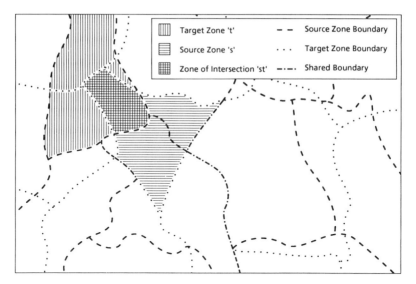

Figure 7.1. 2 source zones, target zones and zones of intersection.

term, coined by Goodchild and Lam (1980), is 'areal interpolation' which has
the advantage of suggesting similarities with other types of interpolation
problem. Reviews are available from Lam (1983) and Flowerdew and
Openshaw (1987).

Terms and notation

First of all, some terminology should be established. The problem of comparing data for two zonal systems can be restated as one of deriving data for one
set of zones given the relevant data for another set. The variable of interest
will be denoted Y; data on Y are available for a set of source zones S but
are needed for a set of target zones T where both S and T cover the same
geographical area. Figure 7.1 illustrates these terms. The value of Y for zone
s is denoted y_s (which is known); the value for zone t is denoted y_t (which is
not known).

If zone s intersects zone t, their boundaries form a zone of intersection st.
Sometimes a source zone will lie entirely within a target zone (or vice versa)
but usually each source zone will be split by the target zone boundaries into
several intersection zones, and each target zone will be similarly split. Usually
there will be a simple relationship between the value y_t for target zone t and
the values y_{st} for the intersection zones into which it is split. The problem of
finding the values for the target zones then reduces to the problem of finding
the intersection zone values.

The size of source, target and intersection zones is highly relevant to the
areal interpolation problem. The areas of these zones will be denoted as A_s,
A_t and A_{st} respectively.

The variable of interest Y may be measured in several different ways. An important distinction can be made between extensive and intensive variables. The terms extensive and intensive are taken from Goodchild and Lam (1980). The variable Y is extensive if its value for a target zone t is equal to the sum of its values for each zone st which intersects zone t:

$$y_t = \sum_s y_{st}$$

Y will usually be extensive if it is a count or if it is a total of some sort (e.g. biomass, customer expenditure, crop production).

The variable Y is intensive if its value for a target zone is a weighted average of its value for the intersection zones. Often the weights will be the areas involved:

$$y_t = \frac{\sum_s y_{st} A_{st}}{\sum_s A_{st}}$$

Proportions, percentages and rates are examples of types of intensive variable. Most interval-scale variables are either intensive or extensive, but some are not: for example, relative relief (the difference between the highest and lowest points in a region) is neither extensive nor intensive.

It may also be important to distinguish between variables according to the type of probability model which could be suggested as having generated them. The most basic distinction here may be between discrete and continuous variables. A count of people or events is likely to follow a binomial or Poisson distribution, perhaps compounded or generalized by some other process, while other variables may be values on a continuous scale. In a very large class of cases, socio-economic data for areal units are likely to be derived from some discrete process. Certain seemingly continuous variables, like percentages or rates, may actually be combinations of counts (e.g. the number of unemployed people divided by the number of economically active people). Other apparently continuous quantities, like the total of consumer shopping expenditure in a shopping centre, or the amount of wheat grown in a parish, can be regarded as a sum of a discrete number of continuously distributed quantities, and hence best modelled as a generalized Poisson process.

The approach to areal interpolation to be developed in this chapter involves the use of additional information about the target zones to help in the interpolation process. This requires the use of at least one ancillary variable available for the target zones; this will be denoted X_t. In some cases, X_t may be a vector of ancillary variables, not just a single variable.

Areal weighting

The standard method for areal interpolation was stated by Markoff and Shapiro (1973) and discussed in detail by Goodchild and Lam (1980).

Because it is based on combining source zone values weighted according to the area of the target zone they make up, it will be referred to here as the areal weighting method. Different forms of the method are applicable for extensive and intensive data.

In the case of an extensive variable, it is assumed that the variable concerned is evenly distributed within the source zone. A subzone constituting half of a source zone will therefore have a value equal to half that of the source zone. In general, the ratio of the value for a subzone to the value of the source zone is assumed equal to the ratio of the area of the subzone to the area of the source zone:

$$\hat{y}_t = \sum_s \frac{A_{st}\, y_s}{A_s}$$

For intensive data, an even distribution within source zones is again assumed. This means that the value for a subzone is the same as the value for the source zone. However, when subzone values are combined to produce an estimate for the target zone, the result is a weighted average of the subzone values, with weights equal to the ratio of subzone area to target zone area:

$$\hat{y}_t = \sum_s \frac{A_{st}\, y_s}{A_t}$$

In both cases, these methods assume that the variable of interest is evenly distributed within the source zones. This is the most reasonable assumption only if nothing is known about their distribution within the source zones. Frequently, of course, we do have additional information about distribution within the source zone, either directly or through knowledge of the distribution of other variables which we would expect to bear some relationship to the variable of interest. It is highly likely, for example, that knowledge of important variables like topography or population distribution would affect our expectations of how other variables might be distributed within source zones. It is the object of our research project to develop more sophisticated versions of the areal weighting method to take account of such additional information.

Areal interpolation using ancillary data

As stated above, our aim is to develop methods of areal interpolation which are more 'intelligent' than areal weighting in the sense that they can take into account other relevant knowledge we may have about the source zones. One form of this, dasymetric mapping, has been known for some time (Wright, 1936) although, as Langford *et al.* (1990) say, it has not been widely used. It is applicable where it is known that part of a zone necessarily has a zero value for some variable. If the variable of interest is population, for example, it may (usually) be assumed to be zero for those parts of a zone covered in

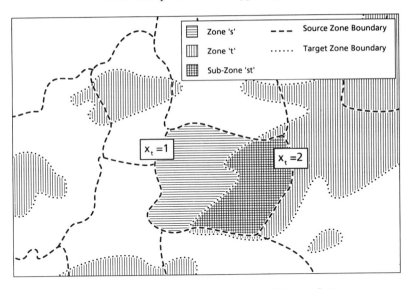

Figure 7.2. Use of ancillary data in areal interpolation.

water, or even for those parts known to have non-residential land use. It is assumed to be evenly distributed within those parts of the source zone not ruled out. This can be regarded as a step towards 'intelligence' but a rather limited one; ancillary information is used to assign a subzone to one of only two categories – either it has the value of the whole zone (or an appropriate proportion thereof) or it has the value zero. The methods to be described here are more flexible in that they allow further information about a zone to be used to make quantitative estimates about the distribution of the variable of interest rather than a crude binary distinction.

The simplest example is where we have ancillary information about a target zone which allows us to assign it to one of two categories. In Figure 7.2, source zone s has a known value for variable Y, y_s, but in order to estimate the value y_t for target zone t we need to know the distribution of y_s between the two subzones created by the intersection of source and target zone boundaries. For example, source zones might be administrative areas for which Y values (e.g. population) are readily available, while target zones might be defined by an environmental feature (e.g. rock outcrop) which might, for historical, economic or other reasons, be expected to support population at a different density. In Southern England, this might be true of limestone and clay belts. If it is desired to know how many people live in a limestone area (where $x_t = 2$ in Figure 7.2), the areal weighting method might give misleading answers if, in source zones like s, the majority of the population is concentrated in the clay portion of the zone.

If it is the case that population density is likely to be higher in a clay area than a limestone area, we would expect that the population of s would be predominantly located in the left part of the zone (dasymetric mapping techniques would not be applicable unless we were sure that nobody lived in the limestone area). Using the same principle for all source zones which include land of both types would lead to the conclusion that the population of a limestone target zone will be less than the areal weighting method would suggest. To estimate how much less, it is necessary to estimate the expected population for a given area of clay (y_1) and for a given area of limestone (y_2). This can be done using the whole data set, as described by Flowerdew and Green (1989), and estimates derived for y_{st} by scaling the results to fit observed y_s values. An essentially similar method has also been applied by Langford *et al.* (1990).

The EM algorithm

This method can be generalised using the EM algorithm (Dempster *et al.*, 1977), a general statistical technique designed primarily to cope with problems of missing data. The EM algorithm consists of two iterated steps, the E step, in which the conditional *expectation* of the missing data is computed, given the model and the observed data, and the M step, in which the model is fitted by maximum likelihood to the 'complete' data set (including the estimates made in the E step). These steps are then repeated until the algorithm converges.

If the areal interpolation problem is considered as one of estimating the values of y_{st} for each subzone formed by the intersection of the source zone and target zone boundaries, it can be regarded as a missing data problem. In general, y_{st} will be missing data but the row totals, y_s, are known. If the y_{st} values can be modelled on the basis of the areas of the subzones A_{st} and ancillary information x_t about the target zones, the EM algorithm can be applied to derive estimates for y_{st}. In practice, of course, y_{st} should only be nonzero for those combinations of source and target zone which actually intersect.

The Poisson and binomial cases

Extending the example in the previous section, where Y represented population, it may be reasonable to assume that Y_{st} has a Poisson distribution with parameter μ_{st}, and that this parameter is a function of x_t, A_{st} and unknown parameters β. The E step then involves computing the conditional expectation $\hat{y}_{st}$ of y_{st} given the data y_s and the current model of μ_{st}:

$$\hat{y}_{st} = E(y_{st} \mid \hat{\mu}_{st}, y_s)$$

$$= \frac{\hat{\mu}_{st}\, y_s}{\sum_k \hat{\mu}_{sk}}.$$

The M step consists of fitting the model:

$$\hat{\mu}_{st} = \mu(\beta, x_t, A_{st})$$

by maximum likelihood as if the estimates y_{st} were independent Poisson data. This will give values for the coefficients giving information about how the Poisson parameters μ_{st} are linked to the ancillary data and the subzone areas. The values of $\hat{\mu}_{st}$ derived in the M step can then be fed back into the E step, to derive better estimates of y_{st}. In turn, these can be used in the M step to provide better estimates of μ_{st}, and so on until convergence is reached. The areal weighting method can be used to give starting values for the y_{st} estimates.

If the ancillary information for target zones is geology, and each target zone is either clay ($x_t = 1$) or limestone ($x_t = 2$), population may be regarded as a Poisson-distributed variable whose parameter is either λ_1 (for clay zones) or λ_2 (for limestone zones). A suitable model for subzone population might be

$$\hat{\mu}_{st} = \lambda_t A_{st}$$

where $t = 1$ if zone t is clay and 2 if it is limestone. The E step might then take the form

$$\hat{y}_{st} = \frac{\lambda_t A_{st} y_s}{\sum_k \lambda_k A_{sk}}$$

for all target zones k intersecting a source zone s. The M step then consists of fitting the model

$$\mu_{st} = \lambda_t A_{st}$$

using $\hat{y}_{st}$ as data. After iteration of the E and M steps until convergence, the target zone totals can be estimated by summing the relevant subzone totals:

$$\hat{y}_t = \sum_s \hat{y}_{st}.$$

This method can be generalized easily for different types of ancillary data within the EM framework. Thus the ancillary variable X could have more than two possible cases; it could be a continuous variable; or there could be several different categorical or continuous variables used in combination to produce improved estimates of y_{st}. Further information is given for these cases in Green (1989).

The models discussed so far are based on the assumption that the variable of interest has a Poisson distribution. This is often a reasonable assumption when dealing with social and demographic data which may be presented as counts of people or events in each zone. In many situations, however, the variable of interest is not the total count, but the proportion of people or events who come into a particular category. We might be interested for example in the proportion of workers in a zone who are unemployed, the proportion of reported crimes using motor vehicles, or the proportion of diseased trees identified in a survey. For problems of this type, a binomial distribution may be appropriate, and Green (1990) has developed methods of areal interpolation applicable to this context.

Again, we assume that data on the proportion of a population sharing some characteristic are available for a set of source zones, but estimates are needed for a set of target zones. In this situation, it is not the area of the subzones produced by the boundary intersections of the source and target zones that is important but their total populations N_{st} (as intersection zones are whole wards, this information can be obtained from the census). We are concerned to estimate y_{st}, the number of 'successes' (i.e. the cases in whose proportion we are interested in), which is regarded as having a binomial distribution with parameters N_{st} and p_{st}, where p_{st} is the probability of a member of the population under study being a 'success'. Ancillary information x_t about the target zones is available and p_{st} can be estimated as a function of x_t and a set of parameters β to be estimated.

The normal case

We have also developed and applied an areal interpolation algorithm suitable for continuous variables (Flowerdew and Green, 1992). We assume that these variables are means of a set of observations in each source zone.

Consider a study region divided into source zones indexed by s. For source zone s, we have n_s observations on a continuous variable with mean y_s. We wish to interpolate these means onto a set of target zones indexed by t.

Consider the intersection of source zone s and target zone t. Let this have area A_{st} and assume that n_{st} of the observations fall in this intersection zone. In practice n_{st} will seldom be known but will, itself, have to be interpolated from n_s. In what follows we may assume that n_{st} has been obtained by areal weighting:

$$n_{st} = \frac{A_{st}\, n_s}{A_s}$$

More sophisticated interpolation of the n_{st} would increase the efficiency of the interpolation of the continuous variable Y but this increase is likely to be small as quite large changes in n_{st} tend to produce only small changes in the interpolated values of Y. We will consider n_{st} as known.

Let y_{st} be the mean of the n_{st} values in the intersection zone st, and further assume that

$$y_{st} \sim N\,(\mu_{st},\ \sigma^2 /\, n_{st}).$$

Now

$$y_s = \sum_t n_{st}\, y_{st} /\, n_s$$

and

$$\begin{bmatrix} y_{st} \\ y_s \end{bmatrix} \sim N\left(\begin{bmatrix} \mu_{st} \\ \mu_s \end{bmatrix}, \begin{bmatrix} \sigma^2/n_{st} & \sigma^2/n_s \\ \sigma^2/n_s & \sigma^2/n_s \end{bmatrix} \right).$$

Clearly if the y_{st} were known, we would obtain y_t, the mean for target zone t, as:

$$y_t = \sum_s n_{st} \, y_{st} / n_t$$

where $n_t = \sum_s n_{st}$.

The simplest method would take $y_{st} = y_s$ to give the areal weighting solution. Here we wish to allow the possibility of using ancillary information on the target zones to improve on the areal weighting method. Adopting the EM algorithm approach, we would then have the following scheme:

E-step:

$$\hat{y}_{st} = E(y_{st} \mid y_s) = \mu_{st} + (y_s - \mu_s)$$

where $\mu_s = \sum_t n_{st} \, \mu_{st} / n_s$.

Thus the pycnophylactic property is satisfied by adjusting the μ_{st} by adding a constant such that they have a weighted mean equal to the observed mean y_s.

M-step:

Treat $\hat{y}_{st}$ as a sample of independent observations with distribution:

$$\hat{y}_{st} \sim N(\mu_{st}, \sigma^2 / n_{st})$$

and fit a model for the μ_{st}

$$\mu^{(s,t)} = X \, \beta$$

by weighted least squares. The notation $\mu^{(s,t)}$ denotes a vector of the elements μ_{st}.

These steps are repeated until convergence and the final step is to obtain the interpolated values y_t as the weighted means of the $\hat{y}_{st}$ from the E-step, i.e.

$$y_t = \sum_s n_{st} \, \hat{y}_{st} / n_t.$$

In practice it is often found that it can take many iterations for this algorithm to converge. Thus although this approach is relatively simple it may not be computationally efficient. However, since we are dealing with linear models we can use a non-iterative scheme as follows. Since ancillary information is defined for target zones only, we have:

$$\mu_{st} = \mu_t, \text{ all s}$$

and

$$\mu^{(t)} = X^{(t)} \, \beta^{(t)}$$

where $\mu^{(t)}$ is the vector of means for target zones and $X^{(t)}$ the corresponding design matrix of ancillary information.

Since

$$\mu_s = \sum_t p_{st} \, \mu_{st}$$

where $p_{st} = n_{st} / n_s$

then

$$\mu^{(s)} = P \, X^{(t)} \, \beta$$

where $P = [p_{st}]$.

Thus we can estimate β directly using data y_s and design matrix $X^{(s)} = P \, X^{(t)}$ by weighted least squares with weights n_s. The final step is to perform the E-step on the fitted values $\hat{\mu}_{st} = \hat{\mu}_t$ computed from $X^{(t)} \hat{\beta}$ and form their weighted means to produce the interpolated values y_t.

Operationalizing the method in GIS

Areal interpolation can be regarded as a fundamental procedure in the use of geographical information systems (GIS) because one of the key operations in GIS is the creation of new areal units using methods like buffering and overlay. Ascribing appropriate attribute values to these new units, as argued above, is highly problematic for non-categorical variables, and hence it is important for a GIS to be linked to methods such as those described above. Ideally, it should be possible for a GIS user to derive appropriate values for new areal units without leaving the GIS; in practice, all that can reasonably be expected is an interface between the GIS and a statistical package which can fit the models described. The statistical package we have used is GLIM (Payne, 1986), which has facilities for interactive modelling for a wide range of statistical distributions, and for adding additional procedures (such as the EM algorithm) in a macro language. In addition to cartographic display, the essential contribution of the GIS is in calculating the areas of the subzones formed by the intersections of the source and target zones. The GIS used in this project was ARC/INFO.

As Kehris (1989) explains, there are problems in linking data between ARC/INFO and GLIM, but an effective method can be developed using the PASS facility in GLIM, provided that the ARC/INFO object code is available. PASS allows GLIM users to import and export data from other software, and its use in this context requires a user-written subroutine to read in an INFO data file. In the example below, the subroutine has been assigned the number 6 which is used to pick out the right code from the PASS facility. The following represents code which can be run from within GLIM to read an item called POPULATION from a polygon attribute table in INFO (in this case, WARDS.PAT in the directory shown) and to put it into a variate called POP in GLIM. The first three lines are GLIM commands and the remainder are generated by the user subroutine (prompts are in italics):

```
? $UNITS 80$
? $DATA POP$
? $PASS 6 POP$
```
ENTER THE DIRECTORY PATH
'/SCRATCH/COV/INFO'
ENTER THE FILENAME
'WARDS.PAT'
ENTER THE USER NAME
'ARC'
ENTER THE NAME OF THE ITEM
'POPULATION'

A similar subroutine may be added to the PASS facility to move data in the reverse direction. Thus source zone data for the variable of interest, target zone data for the ancillary variables and area data for the intersection subzones can all be read from INFO to GLIM. The appropriate version of the areal interpolation algorithm can then be run within GLIM, and the estimated target zone values can be passed back to INFO using the second PASS subroutine for subsequent graphic display and further analysis.

Case studies

The Poisson case

The methods described above for Poisson-distributed variables can be illustrated using 1981 census information for north-west England. Local government districts are the source zones and parliamentary constituencies (using 1983 boundaries) are the target zones. One advantage of using this particular data set is that the data are available for both sets of units; this makes it possible to estimate values for the constituencies from the district data, and then to check the results against the real data. Both districts and constituencies are aggregations of wards. In some cases, districts and constituencies are exactly co-extensive (so areal interpolation is unnecessary); even when this is not so, they often have boundaries which are co-extensive for part of their length. Figure 7.3 shows the boundaries for both sets of units.

There are thousands of count variables available in the Small Area Statistics from the 1981 census for which areal interpolation methods based on the Poisson distribution may be appropriate. The method is illustrated here using the population born in the New Commonwealth or Pakistan (NCP) as the variable of interest. Using districts as source zones, the technique described above was used with several different target zone variables used as ancillary data, some from the census and some from other sources.

Some idea of the effectiveness of the method may be gained from the goodness of fit of the model fitted in the M step of the EM algorithm. This

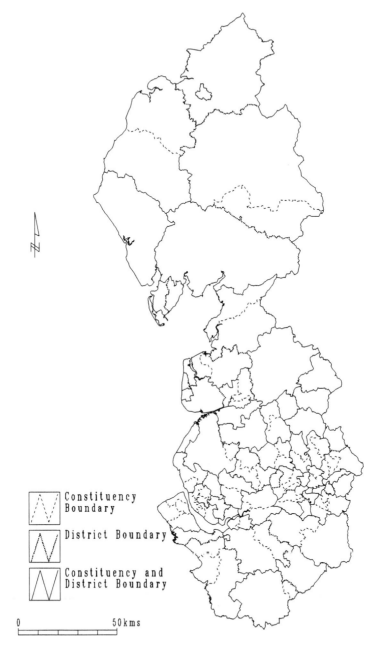

Figure 7.3. Constituency and district boundaries.

Table 7.1. Goodness of fit of areal interpolation models – population born in the New Commonwealth or Pakistan, North-West England, 1981.

Model	Source deviance	Target deviance
Poisson		
Areal weighting	347,834	115,638
Political party	283,515	54,499
Car ownership (categorical)	117,394	62,030
Cars per household	127,868	53,726
Overcrowding	183,240	40,667
Binomial		
Areal weighting	90,651	44,622
Political party	84,933	34,436
Car ownership (categorical)	69,398	34,329
Cars per household	69,912	32,748
Overcrowding	65,145	26,621

is assessed using the deviance (likelihood ratio); lower values represent better model fit. Table 7.1 shows the goodness of fit for five areal interpolation models ('source deviance'). The areal weighting model is the conventional method described by Goodchild and Lam (1980) in which the variable of interest is assumed to be evenly distributed within the source zones. Bearing in mind that the target zones are constituencies, a simple ancillary variable is the political party winning the constituency at the 1983 election (effective use of an ancillary variable in areal interpolation does not require there to be a clear cause-effect relationship with the variable of interest). Car ownership, taken from the 1981 census, is used as an ancillary variable in two different forms, first as a three-level categorical variable and second as a ratio (cars per household). Lastly, overcrowding (proportion of people living at more than one person per room) is used. The figures in the source deviance column are difficult to interpret individually, but a comparison between them suggests that all the other models are considerably better than the areal weighting method, even the rather simplistic political party model. The most successful is the categorical version of the car ownership model, which reduces the deviance to about one-third of its value for the areal weighting model.

Normally this source deviance statistic would be the only way of evaluating the success of the areal interpolation exercise. In this case, however, we know the true target zone values of y_t as well as the source zone values y_s. The estimated values can then be compared with the true values for the target zones; the results of this comparison are given in the 'target deviance' column, which is based on the fit of the estimated y_t values to the known y_t values. The values in this column are comparable with each other but are not comparable with the source deviance values.

Again all the models fitted perform far better than the areal weighting method. However, there seems considerably less difference between the other models than the source deviance figures would suggest. The political party model actually gives better estimates than the car ownership models, and the overcrowding model is the most successful. The implications of this are, first, that 'intelligent' areal interpolation does provide considerably better estimates than areal weighting and, second, that the best model in terms of source deviance is not necessarily the best in terms of target deviance. In addition to the models presented in Table 7.1, further models were fitted including combinations of the ancillary variables described already. Although these models had slightly lower source deviances than the single-variable models, the target deviances did not show similar improvement. Indeed, our experience suggests that a simple model is likely to be more successful than a more complicated one.

The binomial case

The work described so far models the New Commonwealth and Pakistani born population as if its distribution was independent of the population as a whole. It may be more sensible instead to model its distribution with respect to the total population, using the binomial model described above. In such a model, the proportion of NCP born people in the total population is a function of one or more ancillary variables. The lower half of Table 7.1 shows the results of fitting binomial models based on the same ancillary variables, again interpolating from districts to constituencies. The target deviance values are considerably less than for the Poisson models – perhaps not surprisingly since the overall distribution of population is taken into account in these models. Again, the source deviances do not give a perfect indication of the success of the interpolation as measured by target deviance, but the results are more consistent for the binomial models than for the Poisson.

The continuous case

House price data were collected for the borough of Preston in Lancashire during January–March 1990 by sampling property advertisements in local newspapers. The sample used in this exercise included 759 properties, each of which was assigned to one of Preston's 19 wards on the basis of address (their locations are shown in Figure 7.4). Sharoe Green ward contained 170 of these properties, Cadley contained 70, Greyfriars 69, and the others smaller numbers, with under 20 in Preston Rural East (3), Preston Rural West (6) and Central (6). The mean house prices for the wards ranged from £145,825 in Preston Rural West to £34,762 in Park (Figure 7.5). Table 7.2 shows the area, population and mean house price for the wards.

Wards were taken as source zones, with postcode sectors being the target zones. A set of 20 postcode sectors covers the borough of Preston, 6 of them also including land outside the borough (Figure 7.6). In some places, such as

Figure 7.4. Preston wards.

the River Ribble and some major roads, ward boundaries and postcode sector boundaries are identical, but usually they are completely independent.
Ward and postcode sector coverages for Preston were overlaid using ARC/
INFO to determine the zones of intersection and their area. There were
many very small zones, some of which may have been due to differences in
how the same line was digitised in each of the two coverages. The smallest
ones were disregarded up to an area of nine hectares. This left 73 zones of
intersection (Figure 7.7).

Ancillary information about postcode sectors was taken from the June
1986 version of the Central Postcode Directory (CPD), the most recent version available to the British academic community through the ESRC Data
Archive. This source lists all the unit postcodes, and gives a limited amount

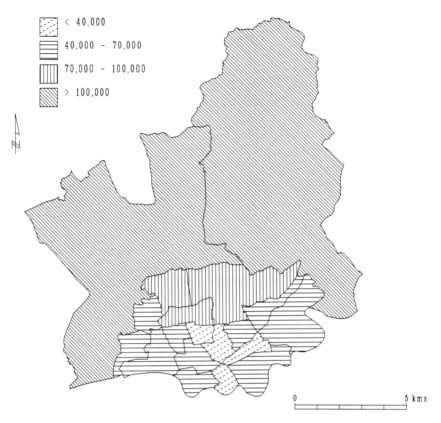

Figure 7.5. Mean house prices for Preston wards.

Table 7.2. Mean house prices in Preston wards (source zones), January–March 1990.

Ward	Population	Area (ha)	Mean house price (£)	Number of cases
Preston Rural East	5,827	5,779	140,667	3
Preston Rural West	9,218	4,597	145,825	6
Greyfriars	6,526	270	91,206	69
Sharoe Green	9,402	800	81,632	170
Brookfield	7,262	176	48,176	21
Ribbleton	6,993	542	54,810	20
Deepdale	6,466	124	47,823	30
Central	4,320	168	40,700	6
Avenham	6,015	149	48,569	29
Fishwick	7,381	322	42,031	39
Park	7,426	134	34,762	25
Moorbrook	5,269	92	38,200	25
Cadley	6,681	193	83,986	70
Tulketh	6,538	101	45,262	58
Ingol	4,564	158	60,294	54
Larches	6,801	177	52,973	26
St Matthew's	6,072	112	37,671	26
Ashton	6,105	276	63,275	54
St John's	5,551	102	35,991	28

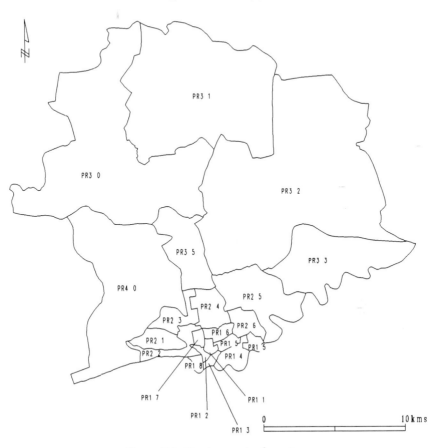

Figure 7.6. Preston postcode sectors.

of information about them: the dates when they were introduced and terminated, and whether they are 'large user' or 'small user' postcodes. Most postcodes in the Preston area have been in existence since the establishment of the CPD in 1981 and are still in use. However, new postcodes are needed in areas of new residential development. Postcodes may go out of use in areas undergoing decline, or in areas where the Post Office has reorganized the system. Large user postcodes usually refer to single establishments which generate a large quantity of mail, such as commercial and major public institutions, while groups of private houses are small users.

The CPD can therefore generate a number of potential ancillary variables, including the number of individual postcodes in a postcode sector, the proportion of postcodes belonging to large users, the proportion of new postcodes and the proportion of obsolete postcodes. Each of these might be expected to reflect aspects of the geography of Preston. Although none of them are directly linked to house prices, they may be correlated with factors which do affect house prices.

Figure 7.7. Preston postcode sectors with wards.

    The most useful one may be the proportion of large user postcodes, which
are likely to be most common in central or commercially developed areas.
The distribution of this variable is shown in Figure 7.8. To some extent, these
areas are likely to have less valuable housing stock, so a negative relationship
to house prices is anticipated. This 'large user' variable was used as an ancillary variable in two forms. A binary variable was created separating those
postcode sectors with more than 10 per cent of postcodes being large users
from those with under 10 per cent large users; and a continuous variable was
created representing the proportion of postcodes belonging to large users.
The latter ranges from .03 in postcode sectors PR2 6 and PR3 0 to .60 in
postcode sector PR1 2.
    Two models were fitted, according to whether the ancillary variable was
treated as binary or continuous. In all cases, the first stage was to estimate
how many sample points were located in each intersection zone. This was
done, as suggested above, by an areal weighting method. The relevant output
from the models comprises the set of estimated house prices for the target

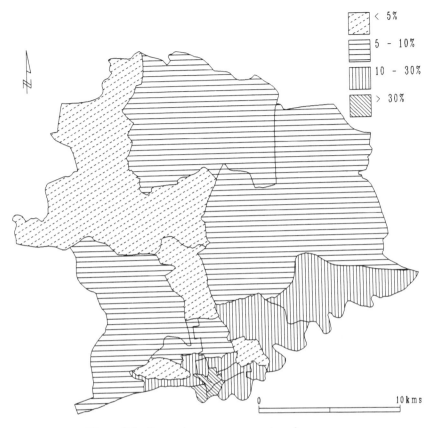

Figure 7.8. Postcode sectors, percentage large users.

zones, the relationship between house price and the ancillary variable, and the goodness of fit of the estimation procedure. The EM and direct estimation methods give identical answers, but the EM method takes longer to reach the solutions – in the binary case, 15 iterations were needed before target zone prices converged to the nearest pound.

Results

Table 7.3 shows the interpolated values for the 20 postcode sectors, together with the ancillary data. It can be seen that there are substantial differences between the areal weighting method (Figure 7.9) and the estimates made using ancillary information (Figure 7.10). There are also major differences between the estimates reached according to the form of the ancillary variable. We do not have information about the true values for the target zones, so it is not possible to evaluate the success of the methods directly. However,

Table 7.3. Interpolated house prices (£) for Preston postcode sectors (target zones)

Postcode sector	Area (ha)	% large users	Areal weighting	Binary variable	Continuous variable
PR1 1	82	44	37,326	35,513	14,578
PR1 2	62	60	41,896	44,592	21,816
PR1 3	68	57	43,554	42,489	20,425
PR1 4	349	23	40,284	38,682	41,322
PR1 5	258	17	43,164	37,784	43,611
PR1 6	175	16	41,216	38,777	44,093
PR1 7	87	13	40,167	40,762	51,096
PR1 8	99	41	48,569	48,569	55,580
PR2 1	413	4	93,164	73,882	69,150
PR2 2	469	27	73,465	44,758	47,295
PR2 3	453	5	91,700	76,745	76,857
PR2 4	567	9	84,618	96,625	87,439
PR2 5	1,725	22	110,793	52,784	66,205
PR2 6	364	3	54,406	75,946	67,160
PR3 0	183	3	145,013	147,321	149,497
PR3 1	223	9	140,667	149,021	140,570
PR3 2	4,437	6	140,781	148,923	144,435
PR3 3	183	11	140,667	106,685	138,045
PR3 5	1,162	4	145,640	147,260	148,283
PR4 0	2,912	6	143,509	97,002	96,692

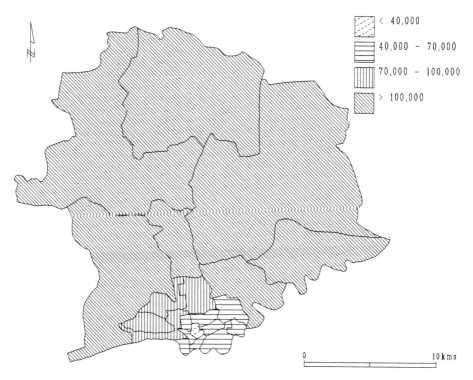

< 40,000

40,000 – 70,000

70,000 – 100,000

> 100,000

0   10 km s

Figure 7.9. Mean house prices interpolated by areal weighting.

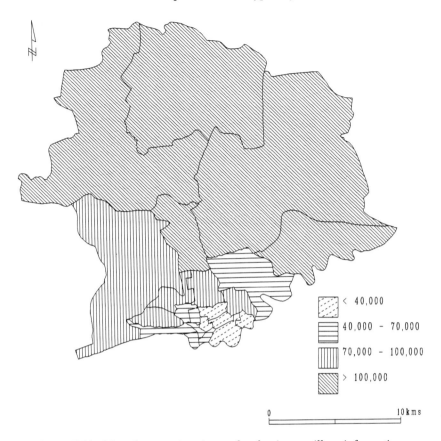

Figure 7.10. Mean house prices interpolated using ancillary information.

the estimated values using the ancillary variable in its continuous form seem unrealistically low for sectors PR1 1, PR1 2 and PR1 3: there are few houses anywhere in England with prices as low as £20,000.

The goodness of fit statistics produced in fitting the models give further information for evaluating them. The house price data are assumed to be normally distributed; a model can therefore be evaluated in terms of the error sum of squares. The total sum of squares for house price values defined over the source zones is 335,600,000,000 (the numbers are very large because prices are expressed in pounds, rather than thousands of pounds, and because they are weighted by the number of houses in each ward).

When the large-user postcode variable is incorporated in the model in binary form, the error sum of squares is reduced to 143,600,000,000. This yields a coefficient of determination (R^2) of .572. The model suggests that postcode sectors with high proportions of large-user postcodes should have mean house prices of £44,056 and sectors with low proportions should have mean house prices of £86,392. The estimated mean values in Table 7.3 are produced by adjusting these figures to meet the pycnophylactic constraint – in other words, ensuring that the observed source zone means are preserved.

It might perhaps have been expected that the crudeness of measurement of the large unit postcode variable would reduce the goodness of fit of the model, and hence that using the proportion of large user postcodes as a continuous variable would improve the model. In fact, however, this model had an error sum of squares of 229,300,000,000, with an R^2 value of only .317. The estimating equation involved a constant term of 84,560 and a coefficient of –126,260. Thus estimated mean house price declined by £126,260 for a unit increase in the proportion of large user postcodes – rephrased in a more easily interpretable way, it declined by £1,263 for a unit increase in the percentage of large user postcodes. As noted above, this resulted in unrealistically low values for those postcode sectors with a high proportion of large user postcodes.

Examination of plots indicates that the relationship of house price to the ancillary variable levels off for higher values and use of a linear relationship can severely underestimate house price for zones with a high proportion of large user postcodes. This is confirmed by incorporating a quadratic term in the relationship which improves the fit (R^2 = .365) and lessens the underestimation problem. Using a logarithmic transformation gives a further small improvement. This highlights the general point that when using continuous ancillary variables careful choice of the form of the relationship may be necessary.

The poorer fit obtained when the ancillary variable was used in continuous rather than binary form is surprising, and suggests the need to experiment with other functional forms. It may also be worth trying other ancillary variables, either singly or in combination. Some other variables may be obtainable through the use of the CPD and digitized postcode sector boundaries, such as the density of small user postcodes, which might be expected to relate to house prices. In many practical applications, variables more directly related to house prices may be available, including of course things like client addresses, provided they are postcoded.

Nevertheless, the improvement in fit of the models discussed in this paper does suggest that taking advantage of ancillary data available for target zones can do much to improve areal interpolation. Without knowledge of mean house prices for postcode sectors, we are unable to quantify the improvement.

Control zones

The methods described so far have been intended to yield improved estimates of variable values for a set of target zones, given ancillary information about the same target zones. There may, however, be situations where target zone estimates may be improved by ancillary information available for a third set of spatial units, which will be referred to here as control zones. This approach was first suggested by Goodchild *et al.* (1989); see also Deichmann *et al.* (1992).

The north-west England study described above was extended by using altitude as a control variable, defining control zones according to whether they were above or below the 400 foot (122 metres) contour, which roughly divides lowland and higher land in the region. Superimposition of source zone, target zone and control zone boundaries produces three-way intersection zones, for example the zone of intersection between Lancaster district, Morecambe constituency and land over 400 feet. The techniques described in previous sections can be generalized for this problem. In practice, altitude made little difference to goodness of fit. The approach might be fruitful, however, with other data or with different control zones.

Difficulties in using intelligent areal interpolation methods

The methods developed in this project represent an improved approach to the areal interpolation problem. They enable additional information to be taken into account when values of a source zone variable are needed for a target zone, and they provide better estimates than the standard method, areal weighting. Nevertheless, there are a number of difficulties in using them.

First, there are practical difficulties in finding the appropriate data and areal unit boundaries and in putting them in the GIS. In the case of this project, there were considerable lengths of boundary which were identical for both district and constituency coverages. However, the two coverages were separately digitized. When one was overlaid on the other, a large number of very small and spurious zones of intersection emerged, created because of small differences in the way the lines had been digitised.

Second, there are problems in linking the GIS used to the analytical routines used in areal interpolation. In our case, we were able to operationalize a link between ARC/INFO and GLIM, but in doing this we were dependent on access to the object code for both packages. It is possible that the link would not be feasible for all hardware platforms, and it may or may not be practicable to construct similar links for other GIS packages or for other appropriate statistical software.

Third, the methods are dependent on finding suitable ancillary variables. The analyst may not have access to suitable target zone data, and the relationship of the variable of interest to available ancillary variables may be tenuous. Our experience has been that some improvement in estimates can often be gained even when the relationship is flimsy, but it is clearly sensible to use ancillary information which is strongly related to the variable of interest, when such information is available.

Fourth, as shown above, there are problems in evaluating the goodness of fit and in comparing models based on different ancillary variables. The source deviance values derived for the models are unreliable guides to the quality of estimates derived as evaluated by target deviance. All we can suggest is that the analyst be guided by a combination of source deviance values, intuitive reasonableness of the relationships postulated and model simplicity. A complicated model with several parameters to be estimated may produce worse target zone estimates than a simpler model based on one ancillary variable, so it may be sensible only to use two or more ancillary variables when there are good theoretical grounds to expect them to have strong and independent relationships to the variable of interest.

Fifth, areal interpolation cannot claim to give totally accurate target zone figures, regardless of what method is used. If interpolated values are incorporated into the GIS, it is desirable that users are aware that they have been interpolated and hence may be inaccurate. Ideally they should be accompanied by some quantitative measure of reliability. This applies not just to the methods described in this paper, but also to the standard methods like areal weighting already used as a default in some geographical information systems. This concern is part of the wider topic of error in GIS and its representation, discussed from many perspectives in Goodchild and Gopal (1989).

Sixth, the areal interpolation methods discussed here are based on a limited set of statistical models, namely Poisson, binomial and normal models. These are appropriate in a wide range of situations, such as analysis of the census variables used in the first case study, but they do not cover all the measurement scales that are likely to be encountered in a GIS. In particular, they do not cover all types of continuous variables, and of course many different continuous distributions must be considered before intelligent areal interpolation methods are generally available for all types of data commonly used in GIS. Work is continuing on extensions of these methods for new types of data.

Acknowledgements

The research on which this paper is based was conducted under research grant R000 231373 from the British Economic and Social Research Council. We are indebted to Evangelos Kehris, who worked on the project, to Susan Lucas for the house price data, to John Denmead and Isobel Naumann for research assistance, to Tony Gatrell and Peter Vincent for ideas and assistance, to the North West Regional Research Laboratory for use of GIS facilities and to the ESRC Data Archive and Manchester Computing Centre for most of the data. Several sections of this chapter are adapted from papers by Flowerdew, Green and Kehris (1991), and Flowerdew and Green (1992).

References

Deichmann, U., Anselin, L. and Goodchild, M. F., 1992, Estimation issues in the areal interpolation of socioeconomic data, in J. Harts, H. F. L. Ottens and H. J. Scholten (Eds.), *EGIS '92: Proceedings of the Third European Conference on Geographical Information Systems*, EGIS Foundation, Utrecht, The Netherlands, **1**, 254–263.

Dempster, A. P., Laird, N. M. and Rubin, D. B., 1977, Maximum likelihood from incomplete data via the EM algorithm, *Journal of the Royal Statistical Society B*, **39**, 1–38.

Flowerdew, R. and Green, M., 1992, Developments in areal interpolation methods and GIS, *Annals of Regional Science*, **26**, 67–78.

Flowerdew, R., Green, M. and Kehris, E., 1991, Using areal interpolation methods in geographic information systems, *Papers in Regional Science*, **70**, 303–315.

Flowerdew, R. and Openshaw, S., 1987, A review of the problems of transferring data from one set of areal units to another incompatible set, *Research Report* 4, Northern Regional Research Laboratory, Lancaster and Newcastle.

Goodchild, M. F., Anselin, L. and Deichmann, U., 1989, A general framework for the spatial interpolation of socioeconomic data, Paper presented at the Thirty-sixth North American Meetings of the Regional Science Association, Santa Barbara, California.

Goodchild, M. F. and Gopal, S. (Eds.), 1989, *The Accuracy of Spatial Databases*, London: Taylor and Francis.

Goodchild, M. F. and Lam, N. S-N., 1980, Areal interpolation: a variant of the traditional spatial problem, *Geo-Processing*, **1**, 297–312.

Green, M., 1989, Statistical methods for areal interpolation: the EM algorithm for count data, *Research Report* 3, North West Regional Research Laboratory, Lancaster.

Green, M., 1990, Statistical models for areal interpolation, in J. Harts, H. F. L. Ottens and H. J. Scholten (Eds.), *EGIS '90: Proceedings of the First European Conference on Geographical Information Systems*, EGIS Foundation, Utrecht, The Netherlands, **1**, pp. 392–399.

Kehris, E., 1989, Interfacing ARC/INFO with GLIM, *Research Report* 5, North West Regional Research Laboratory, Lancaster.

Lam, N. S-N., 1983, Spatial interpolation methods: a review, *American Cartographer*, **10**, 129–149.

Langford, M., Unwin, D. J. and Maguire, D. J., 1990, Generating improved population density maps in an integrated GIS, in J. Harts, H. F. L. Ottens and H. J. Scholten (Eds.), *EGIS '90: Proceedings of the First European Conference on Geographical Information Systems*, EGIS Foundation, Utrecht, The Netherlands, **2**, 651–660.

Markoff, J. and Shapiro, G., 1973, The linkage of data describing overlapping geographical units, *Historical Methods Newsletter*, **7**, 34–46.

Norris, P. and Mounsey, H. M., 1983, Analysing change through time, in D. Rhind (Ed.), *A census user's handbook*, London: Methuen, pp. 265–286.

Payne, C. D. (Ed.), 1986, *The Generalised Linear Interactive Modelling system, Release 3.77*, Numerical Algorithms Group, Oxford.

Walford, N. S., Lane, M. and Shearman, J., 1989, The Rural Areas Database: a geographical information and mapping system for the management and integration of data on rural Britain, *Transactions, Institute of British Geographers*, **14**, 221–230.

Wright, J. K., 1936, A method of mapping densities of population with Cape Cod as an example, *Geographical Review*, **26**, 103–110.

8

Spatial point process modelling in a GIS environment

Anthony Gatrell and Barry Rowlingson

Introduction

Analytical limitations of existing GIS

One of the major functions of a Geographical Information System (GIS) is that of spatial analysis, which we take in general to mean the statistical description or explanation of either locational or attribute information, or both (Goodchild, 1987). However, several authors have bemoaned the lack of analytical power within contemporary GIS. Goodchild (1987, p. 334) observes that 'most contemporary GIS place far more emphasis on efficient data input and retrieval than on sophisticated analysis'. Openshaw *et al.* (1991) seek a set of generic spatial analysis tools for incorporation into GIS, not necessarily the 'sophisticated' ones that Goodchild perhaps envisages. Anselin and Getis (1991, p. 13) note that 'the challenge to the GIS field is that it has not yet been able to furnish or incorporate the types of analytical tools that are needed to answer the questions posed by regional scientists'. For 're-gional scientists' we might substitute geographers, geologists, epidemiologists, environmental scientists, or indeed any user group with an interest in handling spatial data. There seems, then, to be a growing consensus among the aca-demic GIS community that the need to endow GIS with spatial analytical functionality is one of the pressing research needs of the 1990s. In what follows we use the term 'spatial analysis' in the narrow sense of statistical exploration and modelling. For some work on marrying other modelling strategies to GIS we draw attention to Carver's (1991a) research on multi-criteria evaluation.

We may begin to see why current GIS is criticized for its analytical defi-ciency if we follow Anselin and Getis's (1991) view of spatial analysis meth-odology. This suggests the following sequence:

1. selection or sampling of observational units;
2. data manipulation;
3. exploratory data analysis; and
4. confirmatory data analysis.

In the first step we will wish to consider the types of spatial objects (e.g. points or areal units) whose locations and perhaps attributes we intend to analyse. In the second we might want to perform some overlay operations of different coverages in the database, or perhaps transform one set of spatial objects into another class. For instance, we might want to create a set of areal units (Thiessen polygons) around point objects (Goodchild, 1987; Gatrell, 1991). In a third, 'data-driven' (Anselin and Getis, 1991) stage we seek adequate descriptions of our data and try to recognise patterns or structures in the data. This style of exploratory data analysis is now well-known in quantitative geography (Cox and Jones, 1981; Dunn, 1989; Haining, 1990). In the final 'model-driven' stage of statistical spatial analysis we attempt to fit a parsimonious model to data. In this stage we are testing an explicit hypothesis or estimating a formal statistical model.

Many vendors of proprietary systems lay claim to the ability of their products to perform spatial analysis. For many in the GIS vendor community use of this term is restricted to the second component in Anselin and Getis's sequence, the manipulation of objects in the GIS database; perhaps the creation of new classes of object, but not analysis in the sense of seeking to model, or even explore, patterns and relationships in the data. Any vector-based system will, for instance, offer the ability to perform point-in-polygon overlay operations as an 'analytical' feature. However, it is left to users to determine what use they will make of counts of various point objects in particular zones.

Currently, few systems offer any capability of modelling, in a statistical sense, either 'raw' spatial data or that created as a result of geometrical operations. Openshaw (1991, p. 19), for instance, has observed that 'a typical GIS will today contain over 1000 commands but none will be concerned with what would traditionally be regarded as spatial analysis'. This is a slight exaggeration, since, as we note below, there are some products on the market that do permit what Openshaw and others would recognise as 'spatial analysis'. But few systems even offer rudimentary statistical analysis (let alone with any spatial component).

Nature of the link between GIS and spatial analysis

Openshaw (1991), Anselin and Getis (1991), Wise and Haining (1991) and Goodchild et al. (1991) have all discussed the type of link that could be sought between GIS and spatial analysis. Following Goodchild et al. (1991) four strategies may be distinguished:

1. free standing spatial analysis software;
2. 'loose coupling' of proprietary GIS software with statistical software (either proprietary or purpose-built);
3. 'close coupling' of GIS and statistical software; and
4. complete integration of statistical spatial analysis in GIS.

The first of these strategies has generally been adopted in the literature. As Goodchild et al. (1991) note, the best-known example is the Geographical

Analysis Machine (GAM) developed by Openshaw for detecting the locations of leukaemia 'clusters' (Openshaw *et al.*, 1987). The fourth strategy requires a willingness on the part of the vendor community to incorporate a set of spatial analytical modules within their software products; but as the major markets for such products are in the utilities and local government, rather than in the research community, this strategy may be rather unrealistic.

For this reason, most attention now focuses on the question of whether a 'loose coupling' or 'close coupling' is desired. In the former, some kind of interface or reasonably seamless link is sought between a statistical package and a GIS. The two components are, however, kept separate, with data exported from and imported to either of the components in a way that is transparent to the user. In 'close coupling' we might seek either to embed statistical analysis functions into a GIS or to embed a limited set of GIS tools into some analytical software package. Either the spatial analysis modules are written into the command language of the GIS or a conventional statistical software package might be adapted to incorporate GIS functions such as overlay operations. We consider briefly the types of links that others have begun to explore.

Existing approaches to linking spatial analysis and GIS

Perhaps not surprisingly it is raster-based systems that have led the way in adding statistical analysis functions to their suite of tools. Since such systems deal with data modelled as a regular tesselation (typically a square grid) there are computational advantages to be had from such data models when implemented; for example, the area of zones is constant, as are the number of neighbours, with the exception of border cells. One well-known grid-based package, IDRISI (Eastman, 1990) offers several spatial analysis modules. These include: the fitting of trend surfaces; the estimation of an autocorrelation coefficient, which can be generalized to other spatial lags; and simple descriptions of point patterns, including quadrat analysis. In addition, 'standard' statistical functions such as linear regression, are also offered.

Although not a vector-based GIS, in the sense of allowing a full range of GIS functionality (Maguire and Dangermond, 1991), INFOMAP (Bailey, 1990) offers a range of analytical functions (both statistical and mathematical) that few, if any, 'proto-GISs' or advanced mapping packages can match. The data structure is that of a 'flat' data file whose rows are locations and columns attributes. This gives the system the look of a spreadsheet. The provision of a language allows new variables to be created; for instance, a smoothed map is defined by a weighted spatial average. In a purely exploratory sense an autocorrelation function relating to a zonal attribute can be calculated and displayed graphically. In a confirmatory role, users can fit linear models (based on ordinary least squares), so that regression and trend surface analysis are straightforward. The user can switch from statistical analysis in one window to 'map view' in another very fluently and this integration gives the software a real utility that simply is not available in a 'genuine' GIS. As the author implies, we should not be too concerned about whether a system is a GIS or not; what matters is that it is a useful analytical tool (Bailey, 1990, p. 83).

Some of the most imaginative work on spatial statistical analysis in an exploratory sense has emerged from Trinity College, Dublin, where a team led by John Haslett has been exploiting the Apple Macintosh environment in order to develop a powerful system for interactive map analysis (Haslett *et al.*, 1990; Stringer and Haslett, 1991). The work revolves around the notion of 'views' of the data, which might be histograms, scatterplots or spatial displays. The user perhaps specifies a subset of the data that is of interest (for instance, a 'region' in the scatterplot or of the histogram). This is then highlighted on the map image. Additionally, we might want to explore the spatial structure in the data by computing a 'variogram' (Burrough, 1986; Isaaks and Srivistava, 1989) that describes the way in which spatial dependence varies with spatial separation of the observations. The user can 'disaggregate' the contributions made to the variogram estimates and see which pairs of observations on the map contribute most to such estimates. In terms of ease of use for exploratory spatial data analysis the software is invaluable.

An early attempt to forge a 'loose coupling' between a proprietary GIS and statistical analysis was made by Kehris (1990a; 1990b), in work at Lancaster's North West Regional Research Laboratory. This involved both linking ARC/INFO with the statistical modelling package GLIM and also, in a more specialized development, investigating how spatial autocorrelation statistics could be computed by accessing the ARC/INFO data structures. The research was part of a larger project on the development of areal interpolation methods. Areal interpolation is the procedure for transforming data from one set of areal units to another set; the project was concerned with statistical modelling of the variable to be transformed (Flowerdew and Green, 1991). As GLIM was the most appropriate software for modelling it became necessary to transfer data between GLIM and ARC/INFO; this was accomplished by adding subroutines to GLIM's PASS facility which interacted with library functions in INFO. The link to GLIM allowed Vincent (1990) to do some research on modelling the distribution of a bird species in Britain. However, the approach has been criticized by Openshaw, largely because such statistical modelling packages 'were not developed with spatial applications in mind' (Openshaw *et al.*, 1991, p. 788). In our view this matters less than good spatial analytical research. Certainly, the GLIM link allows one to fit a generalised linear regression model, to map residuals (within the GIS) and then to estimate levels of autocorrelation in the residuals.

Ding and Fotheringham (1992) have explored the possibilities of a 'strong coupling', developing a statistical analysis module ('SAM') that runs wholly within the operating environment of ARC/INFO. The software that performs the statistical analysis comprises a set of programs written in C. These are, however, transparent to the user, who simply issues a macro-level command within ARC/INFO (using ARC macro language, or AML) together with appropriate parameters. As with Kehris' work, the current version concentrates on measures of spatial autocorrelation (and also association) and exploits the topological data structures (e.g. the AAT and PAT files) used in ARC/INFO. Related work has been done at the University of Newcastle, where Carver (1991b) has devised AMLs to add error-handling functionality to ARC/INFO.

More recent work at Lancaster University, which we report on briefly below and in more detail elsewhere (Rowlingson and Diggle, 1991), has sought not so much to exploit links between a statistical package and a GIS but rather to add spatial functionality to an existing statistical programming language, S-Plus. Such an approach does not fit comfortably into the fourfold division described above, and although it is one of the strategies reviewed by Openshaw (1991) it is criticized by him as 'an irrelevant distraction' (Openshaw *et al.*, 1991, p. 788). We disagree with this assessment, since it is clear that S-Plus provides a language within which uninhibited exploration of spatial data is quite feasible. For instance, measures of the density of point events ('first-order' descriptions, in spatial statistics parlance) and of the relative locations of pairs of points ('second-order' measures) are easy to implement, as are both views of data and formal tests of space-time interaction. We illustrate some of the possibilities below.

Infusing a proprietary GIS with spatial point process modelling capabilities

As we have seen already, 'spatial analysis ranges from simple description to full-blown model-driven statistical inference' (Anselin and Getis, 1991, p. 5). We begin by describing work that links a proprietary GIS, ARC/INFO, to code for computing descriptions of spatial point patterns (exploratory analysis); next, we describe work that links ARC/INFO to code designed to estimate a spatial point process model (confirmatory analysis). We do not claim to offer a set of generic spatial analysis tools.

Our work is couched within the context of using ARC/INFO for simple reasons of expediency. It happens to be a system with which we have several years of experience at NWRRL; at Lancaster, it runs on SUN SPARC workstations. ARC/INFO is available cheaply to UK academics via the so-called 'CHEST' deal. In addition, we have acquired the object code. Hopefully, this reliance on ARC/INFO is not seen as too restrictive, and is justified by the widespread use of this system in the academic community. Without doubt, our approach could be adapted to suit other proprietary systems.

We outline the generic approach to each of the two problem areas and illustrate the methods with applications. Before doing so, however, we need to describe the type of application that motivates the research and the data environment that makes this possible.

Geographical epidemiology using spatial point process methods

Studies of disease distributions at local and regional scales commence with notifications of disease (morbidity) or death (mortality) among a set of individuals. If we have addresses of such individuals we can map the disease data as a set of 'point events'. Alternatively, we can aggregate the cases to form a set of counts for small areas (which might be Census units, for example). If doing the latter, we usually standardise the counts for age and sex variation. While this is a very common strategy it is not always appreciated that the results are dependent upon the configuration of areal units. As a result, it has become common recently within studies of geographical epidemiology to conduct analyses that use the point events themselves (Bithell, 1990; Selvin *et al.*, 1988; Marshall, 1991). Of course, such analyses must allow for natural background variation in population distribution.

Access to individual addresses raises serious issues of confidentiality. Studies in Britain make use of postcoded data. Since the full unit postcode is shared with, on average, 15–20 households, it offers a sufficiently fine level of resolution to permit some useful analysis over the scale of (for instance) a medium-sized urban area. The link between a unit postcode and a point location is provided by a large machine-readable file known as the Central Postcode Directory (CPD). This gives 100 metre grid references for each of the approximately 1.6 million postcodes in Britain. Further details concerning the quality of the CPD data and their use in epidemiology are given in Gatrell *et al.* (1991a).

Describing the relative locations of pairs of point events

Note that point events within a GIS context are not restricted to epidemiological data. They might represent earthquake epicentres at a global scale, locations of hospitals at a regional scale, or locations of hazardous chemical stores in an urban area. Traditionally, geographers have used methods such as quadrat and nearest neighbour analysis (Haggett *et al.*, 1977) to describe the spatial patterning of such objects. One goal, for instance, has been to assess whether such objects are distributed at random in geographical space. More recently, geographers (e.g. Getis, 1983) have realized that better descriptions of point patterns are to be had by making use of 'second-order' methods, those that describe the relative positioning of pairs of points (Diggle, 1983). One such method is to compute a 'K-function', which describes the relative locations of pairs of point events at different distances. An empirical distribution can be compared with a theoretical distribution derived from a spatial point process model (the simplest of which is the Poisson model of complete spatial randomness).

The estimation of a K-function is implemented via a procedure called **mikhail** ('Mission Impossible K-hat in ARC/INFO at Lancaster'); this is FORTRAN code that calls ARC/INFO routines and accesses a point coverage within the GIS. Output is a series of values of K(s), estimated at distances s. This output can then be displayed graphically using a module, **arcgraph**, developed in ARC/INFO as a link to the UNIGRAPH package.

The form of the command is as follows:

mikhail <pointcov> <polycov> <polyid> <range> <steps> {outfile} {scale}

where: pointcov is the ARC/INFO point coverage; polycov is a polygon coverage (with which the points are associated); polyid is the identifier of that polygon within the polygon coverage that contains the points; range is the maximum distance at which K(s) is estimated; steps represents the number of intervals at which K is estimated; outfile is an optional file if output is not directed to the screen; and scale is another optional parameter for scaling the distances.

The module can be run either from within or outside ARC/INFO. It first reads the ARC/INFO LAB file that stores the {x,y} coordinates. An array of distances for which K is to be estimated is set up; then the K-function estimation is performed within a fixed polygonal region.

To illustrate the K-function estimation we explore some data concerned with gastrointestinal infection ('food poisoning') in Lancaster District. Data were collected between 1985 and 1987 on two such infections (Salmonella and Campylobacter) and included a spatial reference in the form of a unit postcode. These were converted to grid references, as described above.

As we noted earlier, the K-function is a useful description of the point pattern. In the present context we would like to compare the distribution of two point patterns, that for Campylobacter and that for Salmonella. We wish to do this within an area defined by a polygonal region. In the present example, this region approximates to the area of Heysham (part of the wider Lancaster District).

We may do this by running **mikhail** twice, where the command takes the form:

mikhail heycases1 heysham 2 100 10

The parameter heycases1 is the point coverage containing the Campylobacter cases; 'heysham' is the polygonal region of interest; the parameter 2 is the numeric identifier of the selected polygonal region; and the numbers 100 and 10 are, respectively, the maximum limit and interval over which K(s) is calculated.

The output (Figure 8.1) is a set of distances, s, and values of $\hat{K}$(s). This information can then be read into **arcgraph** for plotting (not shown here). In order to compare the points with another point coverage we need to run **mikhail** again and then subtract the two K-functions. A formal test of the difference between the patterns is described and illustrated in Diggle and

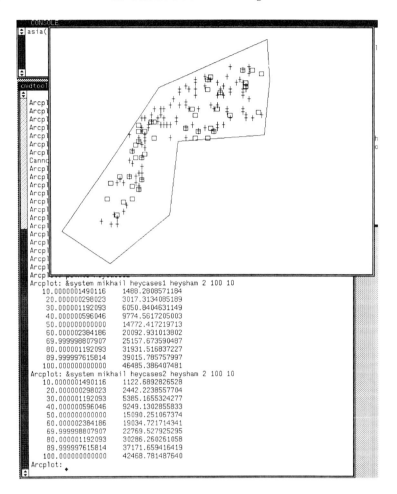

Figure 8.1. Estimation of K-functions in an ARC/INFO environment.

Chetwynd (1991) and in Gatrell *et al.* (1991b). We look at this in more detail within an S-Plus context.

Modelling raised incidence of events around point sources

As noted earlier, Openshaw has been critical of formal hypothesis-testing: '[s]tochastic modelling of spatial pattern is [also] considered an unreasonable objective' (Openshaw *et al.*, 1991, p. 790). We do not share this view. Frequently, scientists wish to test explicit hypotheses derived from prior theoretical work. We do not see why their goals should be excluded from a GIS framework. Nor do we see why a 'data-rich' environment requires us to forego any attempt at testing sensible hypotheses.

This is particularly so in epidemiological work (Thomas, 1991; Marshall, 1991). Our second example of embedding statistical modelling in a GIS framework therefore shows how to estimate a formal spatial statistical model within ARC/INFO. Although the model is a quite specific one, the ideas generalize to other such models.

By way of background, we sometimes require models that assess whether the incidence of rare diseases (such as cancers) is raised in the vicinity of suspected point sources of pollution. A substantial body of work now exists on this topic (see, for example, Hills and Alexander, 1989; Marshall, 1991; Thomas, 1991), driven largely by concerns about raised incidence of childhood leukaemia around nuclear installations. One particular model is of interest to us, that due to Diggle (1990). The model is of the following form:

$$\lambda(\mathbf{x}) = \rho\lambda_o(\mathbf{x}) \; f(d; \; \alpha,\beta)$$

where

$\mathbf{x}$ is a vector of locational coordinates, $\{x_1, x_2\}$; $\lambda(\mathbf{x})$ represents the 'intensity' (local density) of cases of a rare disease; $\lambda_o(\mathbf{x})$ is the intensity of 'controls'; d represents distance from the putative point source of pollution; α and β are parameters of a distance-decay function to be estimated; and ρ is a scaling parameter. The controls are taken to be some measure of background population; or, possibly, locations of a commoner disease not thought to be associated with the putative point source. The null hypothesis is that α and β are zero; the functional form for $f(\cdot)$ is chosen such that it is unity under the null hypothesis. In this case, intensity of the disease cases is simply equal to intensity of the controls, scaled by ρ, and the location of the point source has no effect. The model is fitted using likelihood methods. Maximum likelihood estimates of α and β are generated and the value of the log-likelihood function is compared with its value when α and β are zero. Twice this difference generates a test statistic that may be evaluated as chi-square.

The model relies upon the definition of a smooth surface of the intensity of controls. This is accomplished using 'kernel estimation' techniques (Silverman, 1986; Brunsdon, 1991) which convert the point coverage of controls to a two-dimensional density surface. The form of this density surface depends upon a parameter that controls the degree of smoothness. Large values produce very flat surfaces, while smaller values result in a 'spikier' surface. Clearly, choice of this parameter is important and a large statistical literature on this has emerged. We adopt a method that seeks an optimal value for this constant, but a separate program must be run to obtain an optimal estimate of this smoothing parameter. We return later to this method of kernel estimation.

The raised incidence model is fitted within ARC/INFO via a module called **raisa** (an acronym to partner **mikhail**; it represents work on **rais**ed incidence in an **A**RC/INFO environment). The command line takes the following form:

arc: **raisa** <cases> <controls> <h> <xsource> <ysource> <alpha0> <beta0> <scale>

where: <cases> and <controls> are the point coverages of cases and controls; <h> is the optimal smoothing parameter; <xsource> and <ysource> are the x and y coordinates for the putative point source; <alpha0> and <beta0> are starting values for iterative estimation of a and b; and <scale> is a parameter to scale coordinates.

Note that the user is given control via the keyboard of specifying any location of interest as the putative point source. We have made this a little more user-friendly via a macro written in ARC Macro Language:

arc: &run raisa

which allows interactive location of the putative point source, via the cursor. This allows considerable flexibility if we wish to test the hypothesis of locally raised incidence around a site of interest. However, we note that it is possible to envisage a 'floating' point source, such that a grid of (m × n) locations on the map is evaluated. Thus the model may be fitted mn times and a surface of 'risk' plotted. This is very much related to Openshaw's Geographical Analysis Machine. One improvement we seek to this method is to incorporate options for performing the kernel estimation within an ARC/INFO environment, thus obviating the need for an additional program to be run in advance of raisa.

We may now show how the module can be employed to assess whether cancer of the larynx has raised incidence in the vicinity of a former industrial incinerator in Lancashire. Such raised incidence was demonstrated in earlier work (Diggle *et al.*, 1990), which took data on the much commoner lung cancer as a set of controls. Here, we show how to fit the model in an ARC/INFO environment and how to assess any location as a putative point source.

The following form of command fits the model to the former location of the incinerator:

arc: **raisa** larynx2 lun2 0.15 35508 41390 1. 1. 1000

Here, larynx2 denotes the coverage of point events representing cases of larynx cancer, while lun2 represents a point coverage of lung cancers. The third parameter is the constant denoting the optimal smoothing of the lung cancer data; the fourth and fifth parameters are the x and y coordinates of the putative point source. The remaining parameters are, in order: starting values for α and β, and a constant that permits scaling of the coordinates. A run of raisa generates screen display output (Figure 8.2) that plots both cases and controls and includes estimates of model parameters and a test statistic to evaluate as chi-square. Note from Figure 8.2 that maximum-likelihood values of the two parameters are given, together with the log-likelihood and the log-likelihood for the 'null' model (where $\alpha = \beta = 0$). Twice this difference is displayed and if this exceeds the critical χ^2 value (as it does here) the model gives a statistically significant fit to the data. In other words, cancer of the larynx, relative to the lung cancer controls, is raised around the site of a former industrial waste incinerator. As we noted earlier, this exercise represents 'confirmatory' spatial analysis.

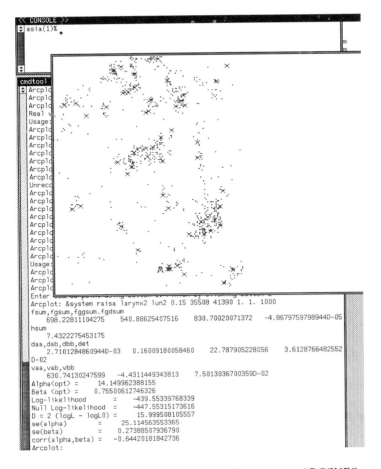

Figure 8.2. Fitting a model of raised incidence of disease in an ARC/INFO environment.

An alternative is to run the model as an AML and to locate the putative point source anywhere in the study region. This uses the model in an essentially exploratory mode. Alternatively, the model may be run over a lattice of locations, generating a grid of parameters and test statistics that may be written to the screen or to a file for further processing (such as contouring to produce a 'risk' surface).

Spatial point process modelling within a statistical programming environment

S-Plus is an interactive environment for statistical data analysis and graphics, building on the earlier S statistical programming language (Becker *et al.*, 1988). A very full range of statistical analysis capabilities is offered, including exploratory data analysis, linear modelling, and so on. Graphical facilities are excellent, including, for instance, the ability to rotate plots in three dimensions. As a tool for the research statistician the language is gaining in popularity.

As a programming language it offers the possibilities of combining basic operations into more elaborate tasks and the definition of new functions. Rowlingson and Diggle (1991) have exploited this to devise a set of routines within S-Plus for the analysis of spatial point patterns. This set of routines is known as SPLANCS.

Among the features offered within SPLANCS are the following:

1. Defining, interactively editing, and displaying point data;
2. investigating point patterns that lie within designated polygonal regions; a point-in-polygon algorithm is incorporated;
3. kernel smoothing (see below);
4. defining empirical distribution functions of point-to-point and 'grid origin-to-point' nearest neighbour distances (G and F-functions, respectively);
5. estimating K-functions;
6. extending K-function estimation to bivariate point patterns; and
7. simulating realisations of stochastic point processes.

To illustrate what is available we can consider first, the example of kernel smoothing and, second, the estimation of K-functions. The latter, in particular, offers the possibility of comparing this approach with that performed within a GIS environment.

We made use of kernel smoothing earlier in fitting the spatial statistical model of locally raised incidence of disease. A SPLANCS function, 'kernel2d', has been devised to do this in S-Plus. Kernel methods are used to obtain spatially smooth estimates of the local intensity of points. The function kernel2d gives the value of $\lambda(\mathbf{x})$ for a specified point pattern, evaluated over a grid of locations that span a particular polygon. The width of the kernel is specified by a parameter. The optimum value of this parameter is derived from another SPLANCS function, mse2d, which uses a method due to Berman and Diggle (1989). The kernel function implemented is a quartic expression:

$$\lambda(\mathbf{x}) = \sum_i (1 - d_i^2/2h_o^2)^2$$

where $\mathbf{x}$ denotes the grid location at which λ is estimated, d_i is the distance from point event i to location $\mathbf{x}$, and h_o is the smoothing parameter.

The surface of $\lambda(\mathbf{x})$ may be displayed in a variety of ways using S-Plus graphics. One useful display is as a raster image; for example (Figure 8.3), we may display the kernel estimates of lung cancers described earlier.

The following code defines K-functions for two spatial point patterns (childhood leukaemias and associated 'controls') in the county of Humberside. It computes the differences between them, and then randomly labels the combined set of points in order to plot upper and lower 'envelopes'. Both map and graphical information may be displayed in different windows, exploiting

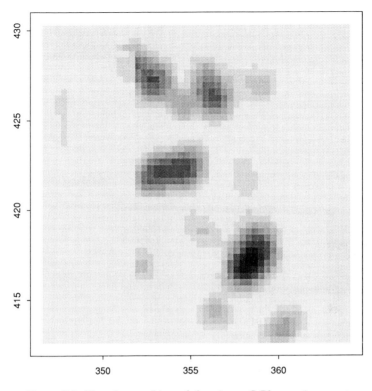

Figure 8.3. Kernel smoothing of data in an S-Plus environment.

the capabilities of a sophisticated windows environment (Figure 8.4). Note that as the difference function lies within the envelope there is no significant difference between the two distributions.

```
> polymap(humberside)
> mpoint(case1,case2,add=T,cpch=c(3,4))
> s <- seq(0,.01,length=21)[2:21]
> k2 <- khat(case2,humberside,s)
> k1 <- khat(case1,humberside,s)
> d12 <- k2-k1
> kenv.cases <- Kenv.label(case1,case2,humberside,50,s)
> matplot(s,cbind(kenv.cases$lower,kenv.cases$upper),col=1,type='1')
> par(lwd=2)
> lines(s,d12)
```

The command polymap plots a polygonal boundary, while mpoint plots the cases and controls. A set of distances is defined in the array, s. The vectors k1 and k2 are assigned values of the K-function, khat, that has been encoded into SPLANCS using FORTRAN code. A new vector, d12, is defined as the difference between the functions; this is plotted as the solid line in the graph

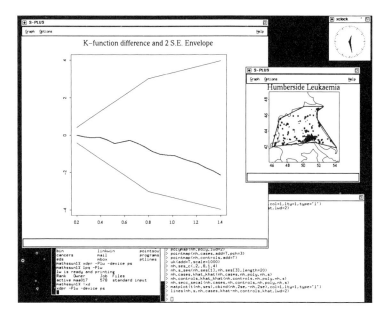

Figure 8.4. Testing for spatial clustering in an S-Plus environment.

window. The function Kenv.label computes the envelope of k1-k2 from a random labelling of the two point patterns and this is displayed using matplot.

Conclusions and further research

Much of what is offered here does not exploit the potential of GIS to examine associations of one set of features (e.g. points) with another (e.g. lines). In effect, what we have accomplished in Section 4 is to do within a GIS environment what spatial statisticians interested in point processes have been doing for 15 years. Attention might now be turned to tasks which are more dependent upon characteristics of the GIS database. One of these might be to examine the locations of points on lines and to define second-order measures on networks. One application concerns the question of whether there are localised aggregations of road traffic accidents (point data) on road networks.

Another problem that requires consideration is that of space-time clustering. Several methods for detecting such clustering are available in the statistical literature (see the review by Williams, 1984). However, a new test for space-time interaction has been developed, one which deals with both space and time in a continuous sense (rather than discretizing space and time in a series of distance and time intervals (Rowlingson and Diggle, 1991)). The method has been applied to data on the distribution of Legionnaire's disease in Scotland (Bhopal, Diggle and Rowlingson, 1992), data that reveal significant space-time interaction.

In this chapter we have followed others in their recent calls for research that aims to infuse GIS with greater spatial analytical power. We hope to have demonstrated some ways forward in accomplishing this task, with reference to both exploratory and confirmatory statistical analysis. We acknowledge that more needs to be done in the way of 'user-friendly' menus and help functions. Nonetheless, the approaches adopted at Lancaster's NWRRL are sufficiently general in scope to be adaptable in a range of spatial statistical analyses.

Acknowledgements

This paper is a revised version of an earlier NWRRL Research Report (No. 23). The research was supported by the Economic and Social Research Council (grant R000 23 2547) and draws heavily on the ideas of Professor Peter Diggle of the Department of Mathematics at Lancaster University. We are indebted to colleagues in the North West Regional Research Laboratory (funded by ESRC grant WA504 28 5005) for their assistance and advice. Credit for the development of SPLANCS belongs entirely to Professor Diggle and Mr Rowlingson and is reported fully in Rowlingson and Diggle (1991). A diskette of the code is available on request, at a cost of £60.

References

Anselin, L. and Getis, A., 1991, Spatial statistical analysis and Geographic Information Systems, Paper presented at Thirty First European Congress of the Regional Science Association, Lisbon, Portugal.

Bailey, T. C., 1990, GIS and simple systems for visual, interactive spatial analysis, *The Cartographic Journal*, **27**, 79–84.

Becker, R. A., Chambers, F. and Wilks, S., 1988, *The New S Language*, New York: Wadsworth and Brooks.

Bhopal, R., Diggle, P. J. and Rowlingson, B. S., 1992, Pinpointing clusters of apparently sporadic Legionnaire's disease, *British Medical Journal*, **304**, 1022–1027.

Bithell, J. F., 1990, An application of density estimation to geographical epidemiology, *Statistics in Medicine*, **9**, 691–701.

Brunsdon, C., 1991, Estimating probability surfaces in GIS: an adaptive technique, pp. 155–164, Harts, J., Ottens, H. F. L. and Scholten, H. J. (Eds.), *Proceedings of the second European Conference on Geographical Information Systems*, Brussels, EGIS Foundation, Utrecht.

Burrough, P., 1986, *Principles of Geographical Information Systems for Land Resources Assessment*, Oxford: Oxford University Press.

Carver, S., 1991a, Integrating multi-criteria evaluation with geographical information systems, *International Journal of Geographical Information Systems*, **5**, 321–339.

Carver, S., 1991b, Adding error functionality to the GIS toolkit, pp. 187–94, Harts, J., Ottens, H. F. L. and Scholten, H. J. (Eds.), *Proceedings of the second European Conference on Geographical Information Systems*, Brussels, EGIS Foundation, Utrecht.

Cox, N. J. and Jones, K., 1981, Exploratory data analysis, pp. 135–143, Wrigley, N. and Bennett, R. J. (Eds.), *Quantitative Geography*, London: Routledge.

162   *A. Gatrell and B. Rowlingson*

Diggle, P. J., 1983, *Statistical Analysis of Spatial Point Patterns*, London: Academic Press.

Diggle, P. J., 1990, A point process modelling approach to raised incidence of a rare phenomenon in the vicinity of a pre-specified point, *Journal of the Royal Statistical Society, Series A*, **153**, 349–362.

Diggle, P. J. and Chetwynd, A. G., 1991, Second-order analysis of spatial clustering for inhomogeneous populations, *Biometrics*, **47**, 1155–1164.

Diggle, P. J., Gatrell, A. C. and Lovett, A. A., 1990, Modelling the prevalence of cancer of the larynx in part of Lancashire: a new methodology for spatial epidemiology, pp. 35–47, Thomas, R. W. (Ed.) *Spatial Epidemiology*, London: Pion.

Ding, Y. and Fotheringham, A. S., 1992, The integration of spatial analysis and GIS, *Computers Environment and Urban Systems*, **16**, 3–19.

Dunn, R., 1989, A dynamic approach to two-variable color mapping, *The American Statistician*, **43**, 245–252.

Eastman, J. R., 1990, IDRISI: A Grid-Based Geographic Analysis System, Department of Geography, Clark University, Worcester, Massachusetts.

Flowerdew, R. and Green, M., 1991, Data integration: statistical methods for transferring data between zonal systems, pp. 38–54, Masser, I. and Blakemore, M. (Eds.), *Handling Geographic Information*, Harlow: Longman.

Gatrell, A. C., 1991, Concepts of space and geographical data, pp. 119–34, Maguire, D. J., Goodchild, M. F. and Rhind, D. W. (Eds.), *Geographical Information Systems: Principles and Applications*, Harlow: Longman.

Gatrell, A. C., Dunn, C. E. and Boyle, P. J., 1991a, The relative utility of the Central Postcode Directory and Pinpoint Address Code in applications of geographical information systems, *Environment and Planning A*, **23**, 1447–1458.

Gatrell, A. C., Mitchell, J. D., Gibson, H. N. and Diggle, P. J., 1991b, Tests for spatial clustering in epidemiology: with special reference to Motor Neurone Disease, Clifford Rose, F. (Ed.) *New Evidence in MND/ALS Research*, London: Smith-Gordon and Company.

Gatrell, A. C. and Dunn, C. E., 1992, GIS and spatial epidemiology: modelling the possible association between cancer of the larynx and incineration in North-West England, de Lepper, M. J. C., Scholten, H. J. and Stern, R. M. (Eds.), *The Added Value of Geographical Information Systems in Public and Environmental Health*, Dordrecht: Kluwer Academic Publishers.

Getis, A., 1983, Second-order analysis of point patterns: the case of Chicago as a multi-center urban region, *The Professional Geographer*, **35**, 73–80.

Goodchild, M., 1987, A spatial analytical perspective on Geographical Information Systems, *International Journal of Geographical Information Systems*, **1**, 327–334.

Goodchild, M. F., Haining, R. and Wise, S., 1991, Integrating GIS and spatial data analysis: problems and possibilities, *International Journal of Geographical Information Systems*, **6**, 407–23.

Haggett, P., Cliff, A. D. and Frey, A. E., 1977, *Locational Methods in Human Geography*, London: Edward Arnold.

Haining, R., 1990, *Spatial Data Analysis in the Social and Environmental Sciences*, Cambridge: Cambridge University Press.

Haslett, J., Willis, G. and Unwin, A., 1990, SPIDER – an interactive statistical tool for the analysis of spatially-distributed data, *International Journal of Geographical Information Systems*, **4**, 285–96.

Hills, M. and Alexander, F., 1989, Statistical methods used in assessing the risk of disease near a source of possible environmental pollution: a review, *Journal of the Royal Statistical Society, Series A*, **152**, 353–363.

Isaaks, E. H. and Srivistava, R. M., 1989, *An Introduction to Applied Geostatistics*, Oxford: Oxford University Press.

Kehris, E., 1990a, A geographical modelling environment built around ARC/INFO, North West Regional Research Laboratory, Research Report 13, Lancaster University.

Kehris, E., 1990b, Spatial autocorrelation statistics in ARC/INFO, North West Regional Research Laboratory, Research Report 16, Lancaster University.

Maguire, D. J. and Dangermond, J., 1991, The functionality of GIS, pp. 319–35, Maguire, D. J., Goodchild, M. F. and Rhind, D. W. (Eds.), *Geographical Information Systems: Principles and Applications*, Harlow: Longman.

Marshall, R. J., 1991, A review of methods for the statistical analysis of spatial patterns of disease, *Journal of the Royal Statistical Society, Series A*, **154**, 421–444.

Openshaw, S., 1990, Spatial analysis and geographical information systems: a review of progress and possibilities, pp. 153–163, Scholten, H. J. and Stilwell, J. C. H. (Eds.), *Geographical Information Systems for Urban and Regional Planning*, Dordrecht: Kluwer Academic Publishers.

Openshaw, S., 1991, A spatial analysis research agenda, pp. 18–37, Masser, I. and Blakemore, M. (Eds.), *Handling Geographical Information*, London: Longman.

Openshaw, S., Brunsdon, C. and Charlton, M., 1991, A Spatial Analysis Toolkit for GIS, pp. 788–796, Harts, J., Ottens, H. F. L. and Scholten, H. J. (Eds.), *Proceedings of the second European Conference on Geographical Information Systems*, EGIS Foundation, Utrecht, The Netherlands.

Rowlingson, B. and Diggle, P. J., 1991, SPLANCS: Spatial point pattern analysis code in S-Plus, *Computers and Geosciences*, **19**, 627–55.

Selvin, S., Merrill, D., Schulman, J., Sacks, S., Bedell, L. and Wong, L., 1988, Transformations of maps to investigate clusters of disease, *Social Science and Medicine*, **26**, 215–221.

Silverman, B. W., 1986, *Density Estimation for Statistics and Data Analysis*, London: Chapman and Hall.

Stringer, P. and Haslett, J., 1991, Exploratory, interactive analysis of spatial data: an illustration in the area of health inequalities, Paper delivered at AGI Meeting, Birmingham.

Thomas, R. W., 1991, Quantitative methods: clines, hot-spots and cancer clusters, *Progress in Human Geography*, **15**, 444–455.

Vincent, P. J., 1990, Modelling binary maps using ARC/INFO and GLIM, pp. 1108–1116 in Harts, J., Ottens, H. F. L. and Scholten, H. J. (Eds.), *Proceedings of the first European Conference on Geographical Information Systems*, EGIS Foundation, Utrecht, The Netherlands.

Williams, G. W., 1984, Time-space clustering of disease, pp. 167–227, Cornell, R. G. (Ed.), *Statistical Methods for Cancer Studies*, New York: Marcel Dekker.

Wise, S. and Haining, R., 1991, The role of spatial analysis in Geographical Information Systems, Paper delivered at AGI Meeting, Birmingham.

9

Object oriented spatial analysis

Bruce A. Ralston

Introduction

Geographic information systems are powerful tools, but their power is in many ways limited. Procedures such as buffering, overlay, and adjacency analysis enhance our ability to perform important tasks. However, there are many methods of spatial analysis which, at present, are difficult to use in a GIS. Indeed, a wide variety of modelling techniques and model structures are yet to be integrated with commercial GIS packages. In this chapter we will present an approach which facilitates the integration of spatial analysis tools with GIS packages, and which can be adapted to a large number of spatial analysis procedures.[1] The basic paradigm used is that of object oriented programming (Entsminger, 1990).

Spatial analysis is not a well defined term, and can consequently embrace a wide variety of techniques and modes of analysis. For example, overlay maps of rainfall, soil type, and crop yield have been used to study the carrying capacity of regions in the Sahel (Moore *et al.*, 1991). While this form of modelling is important, it is not the focus of this paper. Here we are concerned with analysis procedures which require the use of mathematical models which lie outside the realm of most, if not all, commercial GIS packages. These procedures would include various statistical calculations and mathematical models of spatial processes. There are some decision support systems which do indeed combine elements of GIS with spatial process modelling (Florian and Crainic, 1989; Ralston and Liu, 1990; Ralston and Ray, 1987; Ralston, Liu and Zhang, 1992). However, these packages are problem specific and do not offer much in the way of general modeling of spatial processes.

What is needed are reusable tools which facilitate the formulation and use of various spatial models. The tools to be developed need to be sufficiently general to be of use in a wide variety of problems, and flexible enough to be tailored to specific problems. Further, they must allow the construction of logical entities needed to accurately model spatial behaviour. The tools should allow users to concentrate on the characteristics of and relationships between spatial objects, and not on how the objects are processed by the computer code. Finally, these tools should expedite the association of model results with the GIS databases.

The object oriented programming paradigm lends itself to developing these kinds of tools. Properties of OOP offer the ability to develop classes and procedures which meet the requirements outlined above. In addition, the

GIS primitives of point, line, and polygon clearly lend themselves to encapsulation, the basic idea in OOP. In the following sections we use an object oriented programming language, C++ (Stevens, 1990; Stroustrup, 1987), to develop a set of classes and procedures for handling models of spatial processes which require the formulation of some type of matrix. We then apply these tools to increasingly complex optimization problems. This is followed by illustrating how the results of the analysis can be related back to the GIS entities. To begin, we must first define some terms and concepts in OOP and C++.[2]

Object oriented programming concepts and terms

In developing the tools described above, we will use the following properties of OOP. Key terms and concepts are in ***bold italic***.

The Concept of an Object Class. The class is the basic unit of analysis in object oriented programming. Classes ***encapsulate*** both data and procedures into a single structure. In developing classes, we need to ask, 'What are the objects of analysis, what are their properties, and what do they do?' The properties of an object form its ***data members***, while the behaviours of an object constitute its ***member functions***. An example object would be like the following:

```
class An_Object{           //double slashes indicate comment lines
private:                    //might be public or protected
   An_Object();            //hypothetical constructor function
   ~An_Object();           //hypothetical destructor function
   int    data_member1;    //hypothetical data members
   char   data_member2[10];
      .          .
      .          .
      .          .
   float member_function1(int j);    //hypothetical member functions
   float member_function1(float x);  //overloaded member function
   float member_function2(char * s);
};
```

In order to have classes, we must specify constructor and destructor functions. A constructor will allocate memory for each instance of the class and assign values to its data members. The destructor will free memory after this instance of the class is no longer needed.

Note that the data members and member functions can be declared as either private, public, or protected. This is useful if we want access to an object's data members, but we do not want our application code to inadvertently change a data member or execute a function which can damage our database. For example, an arc attribute file will contain Fnode_, Tnode_,

Polygon_left, Polygon_right, ID, and length values. These values are based on the spatial structure in the data, and we may want to limit an application program's ability to change them. By declaring them as private, we can minimize the chances of applications code inadvertently changing their values.

Suppose, for example, we were studying matrix based models. A reasonable approach would be to develop classes called ROW and COLUMN, because every matrix is composed of rows and columns. Each row might have a Name as one of its data members, and each column might have a Name, too.

What do rows and columns do? That is, what types of procedures might we encapsulate into rows and columns? It seems reasonable to assume that one goal of matrix modeling with row and column classes is to determine the coefficients which will be in a matrix. This, in turn, will be dependent on how a given row and column interact. That is, we will need member functions which will capture the interaction between rows and columns. The interaction will be based on messages passed between rows and columns.

Message Passing. In OOP, the behaviour of objects is dependent on the messages they are passed. In the present example of matrix construction, a column will interact with rows. These interactions will be sensitive to the messages these objects receive. A simple analogy would be to consider a room full of buyers and sellers. A seller might receive a message that a particular buyer has just entered the market. The seller would then send that buyer a message of just what type of product he or she is selling. If it is not something the buyer wants, he or she will ignore the message. If it is something the buyer wants, the buyer and seller will then engage in a negotiation which might lead to a sale.

A similar strategy is used when building matrix coefficients. A column is sent the address of a row which is seeking to find the coefficients which define it. The column will send the row a message. In this case, the column will send its address which tells the row, not only where to find the column, but *just what type of column it is*. If the two interact, a non-zero coefficient may be generated. Otherwise, the coefficient will be zero.

To put this concept in operation, we need to define some terms. **arow** will be a pointer variable which holds the address of a row, and **acol** will hold the address of a column. For each member of the class COLUMN we define the following procedure:

```
float COLUMN::Action(ROW * q)
{ return(q->Reaction(this));}
```

This fragment of C++ code simply states that when a column receives the address of a row (q is a pointer to a row), it sends to that row its (the column's) address. The *this* pointer is used to send the address of a particular instance of the class COLUMN. For each member of the class ROW we define a Reaction procedure as:

```
float ROW::Reaction(COLUMN *p)
{ return(0.0);}
```

This function simply states that the default value of any coefficient is 0.0. Before we can determine how other coefficients will be generated, there are two concepts we must introduce.

Class Inheritance. In OOP, each class can have child classes. In the present example, the ROW and COLUMN classes will serve as base classes. All other classes will be child classes. A child class inherits from its parents all the parent's data members (e.g. Name) and functions (e.g. Actions and Reactions). Suppose for example, we wanted to associate with every arc in a network a variable (column) called ROAD. We could then declare ROAD to be a child of COLUMN. Because it is a child of COLUMN, ROAD would inherit COLUMN's data members and functions. It could also have data members and functions which are unique to it. For example, a ROAD would have a FROM_NODE and a TO_NODE. We might want ROAD to override some of its parent's functions. In particular, we would want ROAD to have its own Action function.

In addition to ROAD, we might want to associate with every node in a region a row called DEMAND. Like ROAD, DEMAND will inherit the data members and functions of its parent, in this case the class ROW. We would want DEMAND to have its own Reaction function, and not use that of its parent.

Virtual Functions. Whenever we want a child's function to always override its parent's action function, we declare that function as virtual. This means that if a function is called, the version which corresponds to the caller's class level is used. If the function is not defined for that level, the definition at the next highest level is used.

Overloading. In C++ it is possible to have a function which can take many different kinds of parameters. The definition of the function may change depending on the parameters which it is sent. This is useful, for example, if a row will react differently to different types of columns. For example, a row corresponding to a storage capacity constraint would react differently to a column corresponding to movements into storage, than to columns corresponding to movements out of storage.

A second type of overloading is operator overloading, which allows operators, such as the =, *, [], + and the casting operator to have different meanings, depending on the arguments on either side of the operator. Thus, with operator overloading, it would be possible to relate one type of object to another simply by saying

```
OBJECT1 = OBJECT2;
```

Core Code. Using the properties of OOP listed above, we can construct what we will call the core code. This code is constructed to model the interaction between objects in the most generic way. That is, it will be written in terms of the base classes. The important thing to note about the core code is that *it will seldom change as the base classes and relationships are applied in new problem settings*. The idea behind the core code is to write general procedures of how objects are processed. That is, to use the base classes whenever possible, and to avoid using the child classes. If constructed properly, the core code procedures will apply to a wide variety of model types and situations.

In the present context, we would define the base classes and their functions as follows:

```
class COLUMN{
private:
        COLUMN    *Next;         //Data member, points to next column
        char          Name[9];      //Data member, column name
        COLUMN() {strcpy(Name," ");  Next = NULL;}   //constructor
        virtual ~COLUMN(){};      //Destructor
        virtual float Action(class ROW *q);   //Function prototype
};
class ROW{
private:
        ROW           *Next;         //Data member, points to next row
        char          Name[9];      //Data member, row name
        ROW() {strcpy(Name," ");  Next = NULL;}   //constructor
        virtual ~ROW(){};           //Destructor
        virtual float Reaction(COLUMN *p) {return 0.0;}   //Reaction
                                                          //function
};
float COLUMN::Action(ROW *q)      //Action function definition
{ return(q->Reaction(this)); }
```

A program to generate a matrix of coefficients would include the following lines of core code:

```
ROW *row, *rtop;   //pointers to the base class, and top of base class list
COLUMN *col, *ctop;   //they also point to the child classes

col = ctop;
while (col)
        {
        row = rtop;
        while (row)
                {
                col->Action(row);
                row = row->Next;
                }
        col = col->Next;
        }
```

This core code generates all the coefficients for a given matrix. That is, as the while loops are processed, every row interacts with every column. More importantly, it does not matter what type of matrix we are constructing, an LP tableau, an adjacency matrix, or an arc-incidence matrix. The core code will not change.

Having defined the core code, the implementation of the OOP approach will depend on the actual problem being studied. That is, it will depend on the definitions of the child classes and how they reflect the attributes and behaviours of the objects being processed. In the next sections, we present a transportation problem with increasing levels of complexity.

Generic LP classes and core code

Suppose we wished to generate an MPS structure file for an LP problem.[3] We would need to augment some of the definitions used above. With every column in an LP we will associate a new data member called COST. With every row, we can associate a right-hand-side and a direction (less than, greater than, equal to, or N). To incorporate these changes, we could either create classes called LPROW and LPCOLUMN which are children of ROW and COLUMN, respectively, or we could alter slightly the definitions of our base classes. We use the latter approach here.

```
class COLUMN{
private:
        COLUMN      *Next;
        char        Name[9];
        float       Cost;              //New data member
        COLUMN() {strcpy(Name," ");  Next = NULL;}
        virtual ~COLUMN(){};
        virtual float Action(class ROW *q);
};

class ROW{
private:
        ROW         *Next;
        char        Name[9];
        char        Direction;         //New data member
        float       Rhs;               //New data member
        ROW() {strcpy(Name," "); Direction = ' '; Rhs = 0.0; Next =
           NULL;}
        virtual ~ROW(){};
        virtual float Reaction(COLUMN *p) {return 0.0;}
        void Write_Row();              //New function prototype
        void Write_Rhs();              //New function prototype
};
```

The new functions in class ROW, Write_Row and Write_Rhs, are used to write the row and rhs sections of the MPS file. The first function writes the Direction for a row and the row name. The second function writes the row name and the right-hand-side value of the constraint, if it is non-zero. The details of their implementation are omitted for brevity.

We would change the core code by adding the following lines (changes are in **bold** type):

```
ROW *row, *rtop; //pointers to the base class, and top of base class list
COLUMN *col, *ctop; //they also point to the child classes
float      coefficient_value;

row = rtop;
while (row)
        {
        row->Write_Row();
        row = row->Next;
        }
col = ctop;
while (col)
        {
        row = rtop;
        while (row)
                {
                coefficient_value = col->Action(row);
                if (coefficient_value != 0.0)
                Write_Line(col, row, value);
                row = row->Next;
                }
        col = col->Next;
        }
row = rtop;
while(row)
        {
        row->Write_Rhs();
        row = row->Next;
        }
```

The function Write_Line will write the column name, row name, and co-efficient value for any matrix element which is non-zero. Its full implementation is omitted for brevity.

A transportation example – part 1

Suppose we have a line coverage which consists of a transportation network, and a point coverage with each point in the coverage corresponding to the from and to nodes in the transportation network. Further, the attribute table for the line coverage contains a field called cost, and the node coverage attribute table contains a field called **demand**. These fields contain the link costs and node demands (or supplies, if negative) for a problem to be solved as an LP on a transportation network. We can then construct the following column and row children.

```
class ARC:public COLUMN{
public:
        int Fnode_, Tnode_;
        ARC(Line *p);                    //constructor prototype
        ARC(Line *p, int j);             //overloaded constructor prototype
        virtual ~ARC(){};
        virtual float Action(ROW *q){return q->Reaction(this);}
}
class DEMAND:public ROW{
public:
        int Place;
        DEMAND (Point *q);               //constructor prototype
        virtual ~DEMAND(){};
        virtual float Reaction(ARC *);   //virtual, overloaded Reaction
                                            function prototype
}
float DEMAND::Reaction(ARC *p)
{
        if (Place == Fnode_)
                return(-1.0);
        if (Place == Tnode_)
                return(1.0);
        return(0.0);
}
```

The constructors take as arguments pointers to the GIS primitives point and line. These pointers will allow the child classes to access the relevant attributes in the attribute table. For example, the DEMAND constructor might consist of the following lines:

```
DEMAND::DEMAND(Point *q):ROW()
{
        Place = q->Id_number;
        Rhs = q->Demand;
        Direction = 'G';
        sprintf(Name,"Z%07d",Nrows++);
}
```

In this example, the DEMAND constraint takes the point's Id_number as Place, and its Rhs from the node's demand. The direction is 'greater than or equal to', since we are dealing with generalized demand constraints. Finally, a global counter, Nrows, is used to construct the row name. The constructors for the class ARC would take the lines Fnode_and Tnode_for its similarly named variables. The cost also would come from the AAT file. The use of the overload in the ARC class constructors is so that twin arcs for two-way flows could be constructed.

The most interesting functions are the Action and Reaction functions. The Action function in class COLUMN is declared as virtual. Thus, when a pointer to a column class calls the Action function (col->Action(row)), the definition of that function is dependent on the type of child function to which col points. The same is true of the Reaction function. When it is called, its definition will be dependent on the type of row the pointer row references.

There is something more involved here. Suppose we had many different kinds of columns, not just ARC. If a Reaction function for the row was not defined for the type of column it was being passed, the virtual property would cause the program to use the closest past generation's Reaction function which is defined for a row. If we always define the default Reaction function for each type of column to be zero, then a coefficient will never be generated for a column which has no relationship with a row. This is how set theory notions such as defining a constraint over a set 'i ∈ A' are implemented. By not defining a function for the class we prohibit it from reacting to an 'inappropriate' message.

There is one other class of functions necessary to implement an LP generator. We need to define procedures for creating the appropriate rows and columns from the GIS primitives of point and line. Perhaps the most flexible method is to use a function which processes each member of the AAT and PAT files. Since PC Arc/Info uses dBase as an attribute file manager, we can use programming tools, such as CODEBASE++ (Sequiter Software, 1991), to access and cycle through the appropriate AAT and PAT files. For example, we might define a function called Assign_Nodes to generate the appropriate Demand constraints.

```
void Assign_Nodes()
{                       //row_bottom points to bottom of row linked list
    for (PAT.top(); !PAT.eof(); PAT.skip())
        {row_bottom = row_bottom->Next = new DEMAND(r);
        }
}
```

A similar function could be defined for creating columns from arcs.

```
void Assign_Lines()
{                       //col_bottom points to bottom of column linked list
    for (AAT.top(); !AAT.eof(); AAT.skip())
        {col_bottom = col_bottom->Next = new ARC(s);
        col_bottom = col_bottom->Next = new ARC(s,1);
                                        //for twin arcs
        }
}
```

A transportation example – Part 2

The previous section outlined how to create an MPS file for a very simple LP problem. In this section, we extend the problem by adding more complexity to the model. To do this, we will need to add new child classes and change the Assign_Points and Assign_Lines functions.

Suppose each record in the AAT table contained information on arc capacities. A non-negative value would indicate that the arc had a known capacity. A negative value would indicate that the arc had infinite capacity, or that its capacity was not an issue. Thus, there would be two types of arcs, those which have a constraint on their capacity, and those which do not. For arcs with capacity limits, an arc capacity constraint would need to be constructed.

To introduce this new wrinkle, we would define the following classes:

```
class CAP_ARC:public ARC{   //CAP_ARC is a child of ARC and is a
                            //COLUMN
{
public:
        float    Capacity;
        CAP_ARC(Line *p):ARC(p){Capacity = p->capacity;}
                                                //constructor
        CAP_ARC(Line *p, int j):ARC(p, j){Capacity = p->capacity;}
                                                //overloaded
        virtual ~CAP_ARC(){};    //destructor
        virtual float Action(ROW *q){return q->Reaction(this);}
}

class ARC_CAP:public ROW{   //an arc capacity constraint
public:
        ARC_CAP(Line *p);        //constructor prototype
        ARC_CAP(Line *p, int j);     //prototype for twin arcs capacity
                                     //constraint
        virtual ~ARC_CAP(){};
        virtual float Reaction(CAP_ARC *p);     //prototype for reaction
                                                //function
}
```

The prototype functions will be similar to those found in previous classes. For example, the Reaction function for ARC_CAP will receive a pointer to a column. If the column is not a capacitated arc, the default function in the parent (ROW) will be used, and a zero will be returned. If the column being considered is a CAP_ARC, then the function will check if it is the arc for which this constraint was constructed. Since each row has a Name, we could encode into the name the arc from which it was constructed. The same could be done for the column. Then, by checking names, we could check if their should be a coefficient (1.0) at this row–column intersection.

The other major change is in the functions which assign points and lines to rows and columns. In this case, we would change the function which assigns lines:

```
void Assign_Lines()
{
    for (AAT.top(); !AAT.eof(); AAT.skip())
        {if (s->capacity < 0.0      // if no capacity constraint needed
            {col_bottom = col_bottom->Next = new ARC(s);
            col_bottom = col_bottom->Next = new ARC(s, 1);
            }
        else                        //capacity arc and constraint needed
            {col_bottom = col_bottom->Next = new_CAP_ARC(s);
            col_bottom = col_bottom->Next = new_CAP_ARC(s, 1);
            row_bottom = row_bottom->Next = new_ARC_CAP(s);
            row_bottom = row_bottom->Next = new_ARC_CAP(s, 1);
            }
        }
}
```

Note that the core code does not need to change.

We can now consider adding more complexity. Suppose we had several modes available, and that each mode was designated by a field in the AAT file. The major changes would be in the data elements for ARCS, CAP_ARCS, and ARC_CAPS. We might want to add a data element called Mode. We could then alter the constructor functions for these classes to assign the mode value from a line's AAT record to the Mode for each class. In this case, neither the core code nor the functions which assign lines and points to rows and columns have to change.

If we then wanted to add intermodal transfer arcs, we could construct new columns TRANSFER_OFF and TRANSFER_ON, to handle such moves. Both classes would have From_Mode and To_Mode as data elements, as well as Place. The core code would not have to be changed. However, the functions which assign points to rows and columns would have to change. In particular, at each node we would have to check for every mode available. If there were two or modes available at a node, we would then construct the necessary transfer arcs.

We could introduce even more complexity into the problem. For example, we might introduce loading and unloading arcs. This would allow us to use fixed charges by mode and to consider delays and additional costs at origins, destinations, and mode transfer points. In a multiperiod problem, we might have storage. That is, we could put things into storage in one period and take them out at a later period.

None of these changes would require a change in the core code. However, we would need to derive new child classes and new functions to assign spatial entities to those classes. That is, we would have to focus on how spatial entities give rise to model variables and constraints, how those variables and constraints interact. We would not have to concern ourselves with how the core code processes those entities.

Relating model results to the GIS database

In order to map the results of a model, we need to relate those results to the GIS database. Further, we might wish to construct easy-to-read tables of the LP results. Consider how we might accomplish this in the present example. We could add to the classes of ROW and COLUMN the data elements of Activity and Dual_Price, and Activity and Reduced_Cost, respectively. The code which created the instances of the different rows and columns when generating the LP model could again be used to relate the spatial primitives to the LP variables and constraints. We could then read the LP results and match the names of the variables to the rows and columns. When a match is found, the values of Activity and Dual_Price, or Activity and Reduced_Cost could be assigned to the proper rows and columns.

In order to directly associate the LP results with spatial entities we could use one of two strategies. We could either add new fields to the AAT or PAT tables, or we could create lookup tables to relate the LP results to the spatial entities. Creating lookup tables would be quite easy since we could either encode the IDs of the points, lines, or polygons into variable and constraint names, or make those IDs data members of the ROW and COLUMN classes.

Suppose we wished to create a table which presented the results of the LP. Writing a table which depends only on rows or columns is easy. We simply cycle through the rows or columns and print out the relevant information found in each. However, if we wish to print a table which might contain both primal and dual information (i.e. information from both a row and a column), we would need a mechanism to associate directly the appropriate row with the correct column. We have such a mechanism. The function

```
col->Action(row);
```

If this function returns a non-zero value, we know that the row and column have something in common. In this way, we could print a line which contained the optimal flow of a commodity on an arc and the dual price on that arc's capacity constraint.

An example

Southern Africa usually experiences its rainy season during the southern hemisphere's summer. The rains during the 1991–2 summer were well below normal. Even countries which usually produce a surplus of grain, such as Zimbabwe and South Africa, need to import large quantities of food grains. Given the poor state of many of the economies in the region, much of the food shipped to southern Africa will be food aid.

While many hopeful events recently have taken place in the region (the end of war in Angola, the return of democracy in Zambia, the signing of a cease-fire accord in Mozambique, and the on-going negotiations for a non-racial South Africa), the transportation system outside of South Africa is still poorly developed. If food aid is to be moved expeditiously to areas in need, logistics planning is needed. We have developed object oriented LP and warehouse routing software for food aid transport for USAID (Ralston, Liu and Zhang, 1992). These models can be run either as a stand alone planning package or as tasks to be run in conjunction with Arc/Info. Here we present the object oriented LP driver.

There are several factors involved in the distribution of food aid. Ports of entry must be chosen, and the best modes and routes to be used must be selected. Locations for pre-positioning food aid in storage facilities also must be selected. Finally, shipping, storage, and distribution decisions must be updated as information from the field is sent back to decision makers at the World Food Program and USAID. In addition, information on the location and cost of bottlenecks, how to set prices for acquiring more carrying capacity, and where to add to storage capacity is needed. This information can be generated from the optimal primal and dual variables associated with an LP.

The information on road, rail, waterway, and air link costs and capacities are kept in attribute files for the different mode coverages. Similarly, information on supplies, demands, and storage capabilities for food aid are kept in point coverages, where the points correspond to some subset of the nodes in the transportation network coverages. The question is how to use this information to generate the proper LP and then to relate the results back to the GIS coverages. An important aspect of this problem is that the 'proper LP' changes as obstacles to distribution in different areas arise. Thus, it is important to be able to quickly change the structure of the LP, and to do it in such a way that the LP generator and the report generator need as little re-coding as possible.

Part of the task in developing a suitable LP formulation is to capture the relevant decision variables, or children of COLUMNS, for the distribution problem. Clearly the amounts to be shipped along the arcs in the link coverages are important. Also important are the various movements which can take place at a node. The arcs which correspond to such movements usually are not part of the digitized database. They need to be constructed. The possible movements at a node are depicted in Figure 9.1.

In addition to generating the COLUMNS, the proper objective function and constraints, or children of ROWS, must be developed. For simplicity, we will assume that the objective function will be to minimize costs.[4] This would include the cost of traversing the transport network, loading and unloading vehicles, storage costs for pre-positioned goods, and intermodal transfer costs.

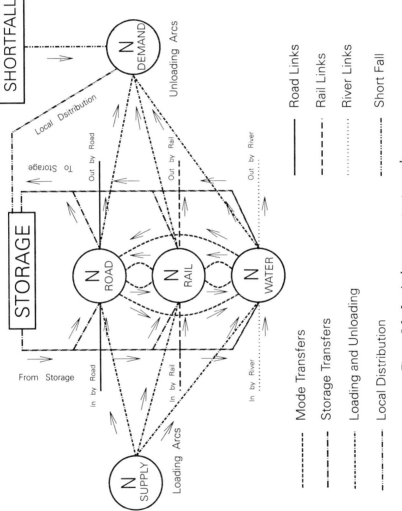

Figure 9.1. *Logical movements at a node.*

We also include a penalty charge for failure to meet the demand at any node. This ensures that all problem formulations are feasible.

The constraints needed to model this problem can be stated as follows:

No arc can carry more than its capacity.

No demand is unmet.

No supply is exceeded.

No storage capacity is exceeded.

Node-by-mode conservation of flow is enforced, i.e. food is neither lost or created at a node.

No port processes more aid than it can handle.

Conservation of storage between time periods is enforced, i.e. food is neither lost nor created at a node from one period to the next.

A complete mathematical statement of the objective function and constraints is given in the Appendix. Here we present the classes needed to create the columns and rows of the LP, indicate their data members, and for constraints, with which variables they interact.

We begin by considering the columns. Every member of the COLUMN family has a constructor function, a virtual destructor function, and a virtual Action function. Here we list the family tree and the elements which differ for each branch.

COLUMN
    Data Members: Name, *Next, Time_Period, Cost;
    ARC
        Data Members: From_Node, To_Node, Mode, Arc_Id
      CAP_ARC
          Data Members: Capacity
    LOADING_ON
        Data Members: Node, Mode
    UN_LOADING
        Data Members: Node, Mode
    MODE_TRANSFER_ARC
        Data Members: Node, From_Mode, To_Mode
    STORAGE_ARC
        Data Members: Node, Capacity
    INTO_STORAGE_BY_MODE
        Data Members: Node, Mode
    OUT_OF_STORAGE_TO_MODE
        Data Members: Node, Mode
    LOCAL_DISTRIBUTION
        Data Members: Node
    SHORTFALL
        Data Members: Node

Every column listed above corresponds to either a physical or a logical arc. The 'family tree' depicted here shows the relationship between each column. The data members for any higher generation are automatically data members of the child classes. In addition, the child classes have data members which are unique to them.

The statement of row classes is necessarily more complex than that for the column classes. It is in the row classes that we articulate how specific rows and columns interact. Usually, this is based on matching similar data members. For example, a coefficient of –1 will be generated if the From_Node of an ARC matches the Node of a Demand_by_Mode constraint, and the Mode and Time_Period match, too.

The family tree for ROWs is:

ROW
        Data Members: Name, RHS, Direction, Time Period, *Next.
        OBJECTIVE FUNCTION
                Data Members: None
                Use: Captures the cost of movement.
                Interaction: Interacts with all children of column.
                                Coefficient is based on cost of column.
        DEMAND
                Data Members: Node
                Use: This constraint makes sure every demand is met.
                Interaction: With UNLOADING and LOCAL
                                DISTRIBUTION
        SUPPLY
                Data Members: Node
                Use: Insures no supplies are exceeded.
                Interaction: With LOADING_ON
        DEMAND_BY_MODE
                Data Members: Node, Mode
                Use: A mode-specific conservation of flow constraint.
                Interaction: With ARC, CAP_ARC, INTO_STORAGE_
                                BY_MODE, OUT_OF_STORAGE_TO_
                                MODE, and MODE_TRANSFER_ARC
        STORAGE BALANCE
                Data Members: Node
                Use: Relates the storage at time t to that at time t-1
                Interaction: With STORAGE_ARC, INTO_STORAGE_
                                BY_MODE, OUT_OF_STORAGE_TO_
                                MODE, and LOCAL_DISTRIBUTION
        STORAGE_CAPACITY
                Data Members: Node
                Use: Keeps storage capacities from being exceeded.
                Interaction: With STORAGE_ARC

ARC_CAPACITY
> Data Members: Arc Id
> Use: Keeps arc capacities from being exceeded.
> Interaction: With CAP_ARC

PORT_OFFTAKES
> Data Members: Node, Mode
> Use: Keeps the flow out of ports, by mode, from exceeding the capacity of the inland transportation system.
> Interaction: With ARC and CAP_ARC

Developing the proper LP structures for food aid logistics has required interaction with personnel of USAID, the World Food Program, and the Southern African Development Coordination Conference. Often, someone will think of an aspect to the problem we have not heretofore considered. This means changing the LP structure by adding new variables or constraints. We have found using the OOP approach to be very useful. It is, in our opinion, easier to adapt existing models to new situations using this approach. The flexibility and adaptability of OOP is key to a high level of productivity and quick turnaround on our part.

Some generalizations

The object oriented programming paradigm used with GIS data types (point, line, and polygon), allows us to develop certain generic tools which can be applied to a variety of problems. Further, they allow us to introduce more complexity with a minimum of new code generation. However, they do more than that. They allow us to concentrate more on the properties and actions of the elements of our problems than on the code needed to generate new problems.

In the example cited above, adding more complexity to the problem required developing new child classes, deciding on their data elements and on their member functions. That is, we had to focus on the elements which define our problem. In developing the child classes we had to consider how rows and columns interact. Since these are generated directly from the spatial primitives, we have to consider how spatial relationships are defined. In some cases, we also had to change the functions which assign points and lines to rows and columns. That is, we have to focus on how spatial entities give rise to new rows and columns in our problem. In this example, this was deciding how nodes and links generate LP rows and columns.

Since the core code will seldom change, and since in a GIS the spatial primitives are of known types (points, lines, and polygons), it should be possible to develop an applications generator which will aid in generating the necessary class definitions. The applications would be similar to query-by-example generators used in database management packages such as dBase IV (Ashton-Tate, 1990). Here we outline how such a generator might work.

Suppose we wished to generate a new child class of ROW. A class generating module would ask the user for the name and parent of the new class.
We could even present the user a family tree, or class hierarchy map, to aid
in choosing the proper parent. Suppose the user's responses were:

name of new class = DEMAND
parent = ROW

We could then present the user with the data elements found in the previous
generations:

parent class has data elements:		
ROW	*Next;	//pointer to next row
char	Name[9];	//name of row
char	Direction;	//New data member
float	Rhs;	//New data member

The user could then add any new data elements by entering their data types
and names (much like constructing a data dictionary in a database program).
In this example, the entry might be

Place Integer;

We could then ask the user to name the type of GIS primitive, point, line, or
polygon, which gives rise to instances of this class. This would be used to
generate how the child's constructor function would pass information to the
parent. In this example, the user would choose point, and we would generate
the constructor prototype:

DEMAND(point *p):ROW(p);

We would have to ask the user how data elements in a point's record define
the data elements in the child class. An example might be:

This class requires a RHS value. From what field in a point's data record
should this be derived?

We could then present the user with a list of field values from which he
could choose. In this case, the choice would be 'demand'.

After all classes are defined, we could then ask how rows interact with
columns. For example, for each child of row, we could present the user with
a list of children of column and ask if a given type of column would interact
with a row. If it did, we could then ask the user to further define how the
interaction takes place.

We have used the OOP approach on other types of problems, as well as
on other types of LPs. One area of particular promise is developing base
classes and core code for algorithms based on adding and substitution strategies. There are many problems which use this approach. For example, a host
of location-allocation problems are based on greedy adding and vertex
substitution algorithms (Teitz and Bart, 1968; Ghosh and McLafferty, 1987).

Warehouse routing problems have also used a variant of the adding and substitution approach (Golden *et al.*, 1977; Clarke and Wright, 1964). Developing classes and core code for these and other generic algorithms has proven useful. Developing procedures for easily generating new applications using these algorithms should make developing spatial analysis applications accessible to a larger audience.

Appendix: the cost minimization model in TRAILMAN

The full statement of the LP model requires the following definitions:

I = the set of arcs, i ∈ I is an index on the set
J = the set of nodes, j ∈ J is an index on the set
J+ = the set of demand nodes, J+ is subset of J
J– = the set of supply nodes, J– is a subset of J
T = the set of time periods, t ∈ T is an index on the set
M = the set of modes, m ∈ M is an index on the set
P = the set of ports, P is a subset of J
S = the set of storage nodes, S is a subset of J
I_{jm-} = the set of arcs i of mode m which leave node j
I_{jm+} = the set of arcs i of mode m which enter node j
AF_{it} is the flow along arc i at time t
AFC_{it} is the per unit cost of AF_{it}
ST_{jt} is the amount in storage at node j at time t
STC_{jt} is the per unit cost of ST_{jt}
TS_{jnmt} is the amount transferred at node j from mode m to mode n at
  time t
TSC_{jnmt} is the per unit cost of TS_{jnmt}
IS_{jmt} is the amount moved into storage at storage node j from mode m
  at time t
ISC_{jmt} is the per unit cost of IS_{jmt}
OS_{jmt} is the amount taken out of storage at storage node j and loaded
  onto mode m at time t
OSC_{jmt} is the per unit cost of OS_{jmt}
LO_{jmt} is the amount loaded onto mode m at supply point j at time t
LOC_{jmt} is the per unit cost of LO_{jmt}
UL_{jmt} is the amount unloaded from mode m at demand point j at
  time t
ULC_{jmt} is the per unit cost of UL_{jmt}
LD_{jt} is the amount removed from storage for local distribution at demand
  and storage node j at time t
LDC_{jt} is the per unit cost of LD_{jt}
SF_{jt} is the shortfall at demand node j at time t

SFC_{jt} is the per unit cost of SF_{jt}

$SCAP_{jt}$ is the storage capacity at storage node j at time t

$ACAP_{it}$ is the capacity of arc i at time t

OFF_{jmt} is the port offtake capacity at port j by mode m at time t

Minimize

$$\sum_{t \in T} \sum_{i \in I} AF_{it} AFC_{it} + \sum_{t \in T} \sum_{j \in J} SST_{jt} STC_{jt} +$$

$$\sum_{t \in T} \sum_{j \in J} \sum_{m \in M} \sum_{n \in M, n < > m} TS_{jnmt} TSC_{jnmt} +$$

$$\sum_{t \in T} \sum_{j \in S} \sum_{m \in M} IS_{jmt} ISC_{jmt} + \sum_{t \in T} \sum_{j \in S} \sum_{m \in M} OS_{jmt} OSC_{jmt} +$$

$$\sum_{t \in T} \sum_{j \in J} + \sum_{m \in M} LO_{jmt} LOC_{jmt} + \sum_{t \in T} \sum_{j \in J^-} \sum_{m \in M} UL_{jmt} ULC_{jmt} +$$

$$\sum_{t \in T} \sum_{j \in J^- \cap S} LD_{jt} LDC_{jt} + \sum_{t \in T} \sum_{j \in J^-} SF_{jt} SFC_{jt}$$

subject to

$$\sum_{m \in M} UN_{jtm} + SF_{jt} + LD_{jt} \geq D_{jt} \text{ for all } t \in T, j \in J^+ \cap S$$

$$\sum_{m \in M} UN_{jtm} + SF_{jt} \geq D_{jt} \text{ for all } t \in T, j \in J^+, j \notin S$$

$$\sum_{m \in M} LO_{jtm} \leq S_{jt} \text{ for all } t \in T, j \in J^-$$

$$\sum_{i \in Ijm^+} AF_{it} - \sum_{i \in Ijm^-} AF_{it} - IS_{jtm} + OS_{jtm} + LO_{jtm} - UN_{jtm} +$$

$$\sum_{n \in M, n < > m} TS_{jnmt} - \sum_{n \in M, n < > m} TS_{jmnt} = 0 \text{ for all } t \in T, j \in J, m \in M$$

$$ST_{jt} = ST_{jt-1} + \sum_{m \in M} IS_{jtm} - \sum_{m \in M} OS_{jtm} - LD_{jt} \text{ for all } j \in S, t \in T, t > 1$$

$$ST_{jt} = \sum_{m \in M} IS_{jtm} \text{ for all } j \in S, t \in T, t = 1$$

$$ST_{jt} \leq SCAP_{jt} \ j \in S, t \in T$$

$$AF_{it} \leq ACAP_{it} \ i \in I, t \in T$$

$$\sum_{i \in Ijm^-} AF_{it} \leq OFF_{jmt} \ j \in P, m \in M, t \in T$$

Notes

1. The examples presented here use terminology associated with Arc/Info. However, they are applicable to many different GIS packages.
2. The next section does not cover all the concepts and properties of OOP and C++. It is meant to highlight those properties which are useful in the present context.
3. An MPS file is the standard input format for a linear program. In an MPS file, only non-zero values of a tableau are stored.
4. In the Transportation and Inland Logistics Manager (Ralston, Liu, and Zhang, 1992), the user can select one of four possible objectives: min. cost, min. time, min. cost and time, and max. flow.

References

Ashton-Tate, 1990, dBase IV Developer's Edition.

Clarke, G. and J. Wright, 1964, Scheduling of vehicles from a central depot to a number of delivery points, *Operations Research*, **12**, 568–581.

Entsminger, Gary, 1990, *The Tao of Objects* (Redwood City, Calif.: M&T Books).

Environmental Systems Research Institute, 1990, PC Arc/Info 3.4D.

Florian, M. and T. G. Crainic, 1989, Strategic planning of freight transportation in Brazil: methodology and applications: Part 1 – project summary, Universite de Montreal, Center de recherche sur les transports, Publication #638.

Golden, B. L., Magnanti, T. L. and Nguyen, H. Q., 1977, Implementing vehicle routing algorithms, *Networks*, **7**, 113–148.

Ghosh, A. and S. McLafferty, 1987, *Location Strategies for Retail and Service Firms* (Lexington, Mass.: Lexington Books).

Moore, D. G., *et al.*, 1991, Geographic modeling of human carrying capacity from rainfed agriculture: Senegal case study, United States Geological Survey, Sioux Falls, SD.

Ralston, B. A. and Liu, C., 1990, The Bangladesh Transportation Modeling System, software package developed for the International Bank for Reconstruction and Development, Washington, DC.

Ralston, B. A., Liu, C. and Zhang, M., 1992, The Transportation and Inland Logistics Manager, software package developed for the Agency for International Development, Famine Early Warning System.

Ralston, B. A. and Ray, J. J., 1987, The Food for Peace Inland Emergency Logistics and Distribution System, software package developed for Agency for International Development, Bureau of Food for Peace and Voluntary Assistance, Washington, DC.

Sequiter Software, 1991, Codebase++.

Stevens, Al, 1990, *Teach Yourself C++* (Portland, Oregon: MIS Press).

Stroustrup, Bjarne, 1987, *The C++ Programming Language* (Reading, Mass.: Addison-Wesley).

Teitz, M. and Bart, P., 1968, Heuristic methods for estimating the generalized vertex median of a weighted graph, *Operations Research*, **16**, 404–417.

PART III

GIS and spatial analysis: applications

10

Urban analysis in a GIS environment: population density modelling using ARC/INFO

Michael Batty and Yichun Xie

Introduction: spatial analysis and GIS

Geographic Information Systems (GIS) technology is spreading rapidly, largely due to the continuing revolution in computer hardware where costs of processing are continuing to drop by an order of magnitude every five years. Workstations are beginning to displace microcomputers for professional computation while mini and mainframe computers are becoming increasingly less useful for GIS applications. At the high performance end, GISs are likely to be ported to mini-supercomputers based on massively parallel computation within the decade while very large applications are being considered on the new generation of supercomputers, or ultracomputers as Bell (1992) has recently called them. All this would not have been possible however without the existence of universally available digital data at the local level. The development of the TIGER files by the Bureau of the Census and USGS now makes possible applications of GIS at many spatial levels, and increasingly data other than that from the Census is being made available to the standards and formats which TIGER requires (Klosterman and Xie, 1992).

In the next decade, GIS will become the backcloth on which most planning and management involving spatial problems in both the public and private domains will take place. In a sense, GIS technology must now be seen as providing the basic framework for processing large volumes of spatial data, and this implies the need to begin to relate and integrate both traditional and new tools for analysis, modelling and design within these frameworks. So far, most applications of GIS stand apart from the battery of tools and methods used by planners and managers to support their decision-making and already it is clear that during the next decade, an intense effort will be needed in adapting GIS to these more traditional established activities. The use of spatial models in physical planning, for example, occurs at each stage of the planning process – analysis, prediction, and prescription (Batty, 1978), and there is the need to develop GIS so that it is useful to this succession of stages. There are also applications of more *ad hoc* techniques of analysis in planning such as those based on various types of indicator which need to be integrated with GIS, while the development of formal methods of spatial decision support based on the optimization modelling is attempting to fuse this technology with techniques originating from management science and operations research (Densham, 1991). In fact, the integration of this whole range of models and design tools with GIS involves creating what Harris (1989) has called 'Planning Support Systems' (PSS). Such systems assume the existence of GIS as the backcloth on which analysis and design would take place in physical and other varieties of spatial planning (Harris and Batty, 1993).

The lack of integration of GIS with planning models and methods simply reflects the state-of-the-art and its history. The emphasis in GIS on data organization and mapping reflects basic operations which are relevant to all spatial analysis and decision-making in that the existence of such data in computable form is an essential prerequisite to use. What spatial functions are present in GIS mainly relate to operations on digital data through overlay analysis and like-methods of combination, reflecting both the ease with which this type of operation can be implemented in such systems and the use of such functions to aid decision-making in the traditional domains of landscape and resource management where GIS first originated. Even such simple functions as associating two variables through correlation are not immediately possible with standard GIS technology although current concern is for providing ways in which existing software incorporating such functions can be easily linked to GIS or purpose-built functions can be integrated through the medium of macro-languages. In fact, to anticipate the work reported in this chapter, our concern will be with the use of such macros to link population density modelling with the data storage and mapping capabilities of state-of-the-art GISs.

A period of intense activity in extending GIS in these terms is already underway. There is much applied work which goes formally unreported in this context while GIS vendors are active in extending their software to embrace such functions, for example through the new network module in ARC/INFO. More specialist GISs which incorporate the functions of specific planning domains such as transport are being developed (for example, TRANSCAD) while in terms of basic research, the National Center for Geographic Information and Analysis (NCGIA) has identified a research agenda for extending the usefulness of GIS to mainstream spatial analysis and modelling (Rogerson and Fotheringham, 1993) of which the papers contained in this book represent the first step. There are other attempts to develop such functionality in GIS in the area of data modelling (Openshaw, 1990) and in spatial process modelling (Wise and Haining, 1991) although at present, a comprehensive assessment of such developments is difficult. There are doubtless many applications in practice which are not formally written up, and there is clearly some inevitable duplication of effort.

We will begin our discussion by exploring the various types of GIS environment in which models and new functions might be integrated. To anticipate our own work, the type of environment in which we will develop our integration is based upon a well-developed GIS – ARC/INFO – in which the GIS itself acts as the organizing frame for a series of external data and model functions. In essence, in this research we are posing two related questions: how far can we use well-developed proprietary GIS such as ARC/INFO to develop analytical spatial models? And what are the limits on such systems with respect to the addition of such functions? In short, we are attempting to find and design ways in which proprietary GIS can embrace such functions without destroying internal consistency or breaking up such software to enable such extensions. There are of course other sides to this question, for example involving the design of models completely separate from proprietary GIS but containing their own purpose-built but generic GIS functions. This is something that we are also working on in the context of the wider project but we consider these issues elsewhere (Batty, 1992).

In this chapter, we will show how ARC/INFO can be consistently extended through its macro-language interface (Arc Macro Language – AML) to link with a variety of external software modules for modelling and data preparation, written in other high-level languages such as Fortran and C. In essence, the GIS will act as the 'display engine' for conventional modelling applications although its use is wider than this in that many of its database management functions are also invoked. We have chosen to demonstrate these issues of model integration using a well-developed proprietary system for several reasons. First as ARC/INFO is well-established, many researchers and professionals have access to it and thus can easily replicate the designs we develop here. Second, ARC/INFO represents our preferred medium for processing digital (TIGER) map files and Census data and thus the data is in this format prior to any analytical work. Third, ARC/INFO is being made increasingly general with respect to its functionality and availability on diverse hardware platforms as well as being popularized through modules such as ArcView. However, it is necessary to note that what we will demonstrate could also be implemented on other proprietary GISs and it is thus the principles of this application which are important.

We have also taken one of the simplest of all spatial models as our point of departure. Population density models are based on relating such density to distance from a central point or pole in the city, usually the CBD, and this requires only population counts for each spatial unit. In fact such models represent the classic merging of attribute and digital data in their definition. The population data is readily available from the 1990 Census down to block group level at present, and it is easily disaggregated into different types. Such models have a strong theoretical rationale, consistent with the conventional micro-economic theory of urban markets dating from Von Thunen to Alonso (1964) and beyond to the new urban economics (Anas, 1987). And they also represent the simplest of spatial interaction models being based on the one-origin, many-destination variant of the standard model (Wilson, 1970), thereby being potentially generalizable to more complex patterns of interaction.

We will begin by sketching pertinent issues involved in modelling within GIS in general and ARC/INFO in particular. We will then change tack and formally state the model and its variants which we are seeking to embed into ARC/INFO. This defines the data requirements and model functions needed and we develop these by outlining the various software modules and how these are integrated. We are then in a position to present our application to the Buffalo region. After some discussion involving our preprocessing of the data, we demonstrate a typical run of the model showing various outputs which act as a visual trace on the user's progress through the suite of programs. As part of these applications, a large number of population density models are estimated and these are reported, notwithstanding their poor performance. These results are not central to this chapter but they do indicate areas for further research which are noted by way of conclusion.

Modelling in GIS environments

There are many different methods for linking spatial models to GIS in general and although there are a lesser number of possibilities for proprietary systems, we need to sketch these to establish the context to our work. We will first examine various types of GIS environment, then present methods of integration, and finally illustrate these notions in a little more detail with ARC/INFO in mind. GISs have mainly been developed from mapping software but for a specific range of applications. However, in the quest to extend to an ever wider array of problem areas, such software has become increasingly general and ever more flexible in its ability to link with other data formats and software. There is little distinction now in terms of their ability to handle raster or vector digital map data and most software is being extended to a range of hardware platforms from PC to supercomputer with the workstation application the main focus at present. More specialized packages have however been evolved recently, for example GRASS by the US Army for resource-based applications, TRANSCAD for transport applications and so on. There also exist many one-off types of application usually designed by large vendors such as IBM for specific problem contexts, some of which are being generalized. In terms of generic GISs, it is possible to use mapping, CAD and spreadsheet software fairly innovatively for GIS-like applications (see Newton, Sharpe and Taylor, 1988; Klosterman, Brail and Bossard, 1993) although most non-standard applications are represented by software which is designed for other purposes and then extended to incorporate more specific and limited GIS functions. To date, there have been few model-based packages which have been extended into GIS although it is likely that there will be many more of these in the next decade as vendors and researchers come to see the need to provide some baseline functionality relating to the processing and display of spatial data whatever the ultimate purpose of the software might be.

With respect to different types of mechanism for linking model-based functions to GIS, there is a range of possibilities from loose- to very strong-coupling. Related to this is the notion that models might be embedded entirely within GIS or GIS embedded entirely within models, usually in the context of strong-coupling. Strategies for loose-coupling are mainly *ad hoc* and simply involve importing or exporting data from separately configured software systems. Strongly-coupled strategies which are favored by the vendors consist of embedding models entirely within the GIS although to date, apart from more specialized systems such as TRANSCAD, only modest progress has been made in this domain. We will not attempt to discuss modelling applications which embed GIS functions entirely within their software here although as already noted, our broader project does concentrate on such possibilities (Batty, 1992).

In fact, the most usual strategies reflect some intermediate position between the loose- and strong-variants. In the tradition to be adopted here, Ding and Fotheringham (1992) have developed spatial autocorrelation functions within ARC/INFO using AML to link with such statistics which are operationalized using C programs. Gatrell (1993) has developed similar strategies linking epidemiological models written in FORTRAN to data in ARC/INFO. Kehris (1990) has linked general linear modelling in the GLIM package to ARC/INFO, while Densham (1993) has developed a variety of user interfaces based on visual interactive modelling through which location-allocation models written in Pascal are interfaced with TRANSCAD, IDRISI and other GISs. There is in fact an explosion of interest in such linkage and it is into this context that our research fits.

It is important too to note the bias towards GIS or external model functions which any such linkage assumes, and this can be quasi-independent of the strength of the coupling. It is possible to have a very strongly-coupled GIS and spatial model where the emphasis is towards the model or towards the GIS. In the application we will describe here, the model is fairly strongly coupled with the GIS through the integrating mechanisms of AML but the emphasis is more towards the GIS than the model in that the emphasis is on using the GIS as the display engine and visual interface. In contrast, in Ding and Fotheringham's (1992) Spatial Analysis Module (SAM) which links spatial autocorrelation statistics to ARC/INFO, the emphasis is more on statistical interpretations and less on the display *per se*. In fact, this emphasis is very much part of the way the interface and integrating mechanisms are designed, and like all design, the ultimate feel of the product to the user is very difficult to gauge in advance if the designer's intuitive abilities and unspecified goals are not known.

In our application, we are using ARC/INFO Version 6 which has fairly elaborate AML functions. The core of our system in fact is ArcPlot which enables maps to be displayed quickly and efficiently, and thus all the functionality required to display and manipulate such maps including overlaying different coverages is invoked as part of the background activity (Huxhold, 1991). AML is really the heart of the system in that these macros control the ArcPlot canvas on which all mapping and graphics linking the data to the model takes place. In fact, ARC/INFO is almost used as a kind of 'paintbox' to illustrate pictures of how the data, the model results and their performance can be judged. Like all pictures, their success lies in the how well the designer communicates the message, and one of the problems posed by proprietary GISs is that such systems have rarely been developed with 'good design' and user-friendly communication in mind. An additional goal of our project therefore is to stretch the limits of ARC/INFO in this regard.

One of the features of AML is the existence of a point and click pop-up menu system from which the entire application can be guided. As we will illustrate later, the entire sequence of stages which involve the calling of various software modules both within and without ARC/INFO is controlled by a main menu system which in turn generates lower level pop-up sub-menus. Some of the functions which we imply are not yet enabled for the project is in its early stages but these will be noted as we progress through the description. The menu system associated with AML however is an example of how such software is becoming more general purpose. This has happened elsewhere, particularly in the design of database management systems and business graphics where spreadsheet, wordprocessing, statistical modules and so on are being incorporated into one another. This is now happening within proprietary GIS and it is likely that many more graphics and other functions at the very least will be added to such systems in the near future as users seek to implement more of their traditional working techniques and analyses within such systems.

Before we digress to explore the types of model to be implemented in ARC/INFO, we should note another feature of our project. Proprietary GISs enable users to handle very large data sets and one of the attractions of developing a graphics interface within ARC/INFO is that the user can be confronted with many opportunities for exploratory spatial data analysis. In terms of the way we have set up the system, the models themselves tend to be appendages to the display routines and thus the emphasis is very much on examining the data before running any of the models. Moreover as a large number of model variants are possible, the emphasis in the project has been upon developing a system in which to explore and estimate models rather than on the model results *per se*. As we will see, the results are not good but less than five per cent of the time spent on the project so far has been on exploring and calibrating the models. The emphasis here has been to shift the focus from the models *per se* to ways in which they can be explored and interpreted, and in this the emphasis on the graphics interface is central. This is but another example of how the existence of GIS software is changing the way we think about and implement analytical and predictive models.

Archetypical modelling problems: population density

Population density models are extremely well-suited to demonstrate the way in which spatial analysis and GIS might be integrated to the mutual benefit of both. We have already noted that population data required for such models is the most standard of all data from the Census and that the most basic link between the digital map and attribute data is through the definition of density. That is, GIS functions are immediately required to calculate areas (in ARC) which in turn are linked to population (in INFO) in deriving the composite variable – density. Furthermore, one of the basic relations of social physics – the monotonic decline of densities with increasing distance from some central economic focus such as a city's CBD – represents the most obvious link between the map component of a GIS and its related attributes in the context of urban applications.

Population density relations are central to urban economic theory. From the work of Von Thunen on, location theorists have demonstrated that the spatial organization of cities is consistent with declining population densities, bid rent surfaces, travel volumes and so on as distance from the CBD increases. Alonso (1964) was the first to show in modern times that such profiles could be derived consistently from micro-economic theory, and this initiated a line of work from Beckmann (1969) to the present which involves the elaboration of the urban micro-economic model and its generalization to multiple markets and spatial scales. Furthermore, with the development of gravity and thence spatial interaction modelling from the 1960s, population density models and their non-normalized equivalents termed by Wilson (1970) residential location models, could be derived using standard optimizing techniques such as maximum entropy, likelihood and variants of utility or welfare, as well as being interpreted as one-origin, many-destination spatial interaction models (Batty, 1974). In this sense then, such models represent a special case of the more general model and their application thus implies such wider generalization.

We will develop two basic population density models here. The first is based on relating density to distance from the CBD as a negative exponential function after Clark (1951) while the second is based on an inverse power function. This latter function has not been widely applied although it now appears that it might have more plausible theoretical properties than the negative exponential (Batty and Kim, 1992). Both of these models will be applied in their normalized form with population density as the dependent variable and in their non-normalized form with population as the dependent variable. There is in fact a third functional form which combines the exponential and the power models in the gamma function (March, 1971) and although this is featured in our software, it is not yet enabled. Finally, we should note that the models we specify are in one-dimensional form in that they refer to the profile of population or population density with increasing distance from the CBD. If the models were specified in terms of orientation as well as distance from the CBD, then aggregating across orientation would give different functional forms than those used here. In future work we will be extending the set of models to the two-dimensional versions based on orientation as well as distance.

We will first state the models in continuous form and then specify their discrete equivalents which are those we will apply. First noting that population density at distance r from the CBD is defined as $\rho(r)$, then

$$\rho(r) = K \exp(-\lambda r), \tag{1}$$

where K is a scaling constant and λ is a friction of distance parameter or in this case a percentage change in density for a small change in distance, at the margin. The model has other important analytic properties relating to its parameters and although these are relevant to its estimation, interested readers are referred elsewhere for this discussion (Bussiere, 1972; Batty, 1974; Batty and Kim, 1992). The second model relates density to an inverse power of distance as

$$\rho(r) = Gr^{-\alpha}, \tag{2}$$

where G is the scaling constant and α is the friction of distance exponent. Lastly the combined model based on a gamma function which we have not yet estimated but which is part of the software can be stated as

$$\rho(r) = Qr^{-\alpha} \exp(-\lambda r). \tag{3}$$

Q is a the scaling constant. The exponents λ and α in equations (1) to (3) are related to the mean density and mean logarithmic density respectively; these are important in estimation using maximum-likelihood and explanations can be found in Wilson (1970), and Angel and Hyman (1976).

We will apply the models in equations (1) and (2) in discrete form and to effect this we need to define population, area, density and distance as follows. The distance from the CBD to zone or tract i is now defined as r_i where it is assumed that $i = 0$ represents the point of origin for which none of the four variables need be defined. Then population in zone i is given as P_i, area of zone i as A_i, and density as ρ_i. Density is defined as

$$\rho_i = \frac{P_i}{A_i}. \tag{4}$$

We must now specify the full set of possible models linking the way the data sets are defined to the various model forms prior to their explicit statement. Before we elaborate these, Figure 10.1 shows a total of 32 model variants based on four characteristics – data sets, model functions, data types and calibration methods. In fact, these four characteristics are ordered according to the way they are selected in the software but we will examine these in a different sequence here. First there is the distinction between functional forms

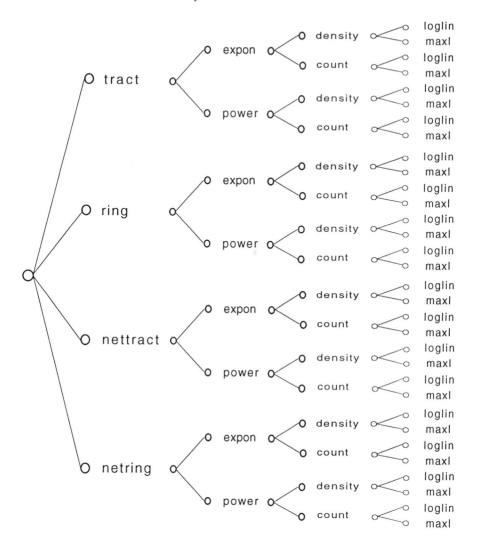

Figure 10.1. Variants of the population models based on different data definitions and model functions.

– negative exponential (abbr: expon) and inverse power (abbr: power) as in equations (1) and (2) above. Second each of these models can be specified in non-normal or normalized form; in the non-normalized form, the dependent variable is population P_i which we refer to as the 'count' data type, while in the normalized form this is density ρ_i, referred to as 'density'.

The third distinction relates to the calibration method used. We have chosen two methods – the first involves logarithmic transformation of each model equation to linear form and then estimation using linear regression (abbr:

loglin), while the second is based on maximum-likelihood (abbr: maxl) which is effected by the solution of the appropriate maximum-likelihood constraint equations (see Batty, 1976). The last characteristic relates to the various data sets defined. We have defined four such sets, namely data based on tracts, mile-wide rings, tracts and residential street networks, and rings and residential street networks: these are abbreviated to tract, ring, nettract and netring respectively. However we will not elaborate the definition of these data sets further at this point. We will simply show how the first three characteristics based on function, data type, and calibration methods are reflected in terms of their relevant model equations. The count and density models based on the negative exponential functional form can now be stated as

$$P_i = K_1 \exp(-\lambda_1 r_i), \quad \text{and} \tag{5}$$

$$\rho_i = \frac{P_i}{A_i} = K_2 \exp(-\lambda_2 r_i), \tag{6}$$

while the inverse power forms of model are

$$P_i = G_1 r_i^{-\alpha_1}, \quad \text{and} \tag{7}$$

$$\rho_i = \frac{P_i}{A_i} = G_1 r_i^{-\alpha_2}, \tag{8}$$

Note that the model parameters are now subscripted as 1 and 2 to distinguish the fact that they refer to count and density estimates respectively. These two sets of models are estimated using loglinear regression or maximum likelihood. Loglinear regression is straightforward with the scaling constants being computed as the exponentials of the relevant regression intercepts and the friction parameters being the slopes of the appropriate regression lines. The maximum likelihood equations which are associated with the scaling and friction parameters of each of these models are trickier to specify. The appropriate equations which are satisfied for each of the four model variants are related to totals and means for the scaling and friction parameters respectively. These are given in Table 10.1 where in the case of the maximum likelihood estimates, the scaling and friction parameters have to be chosen so that each of their associated equations is satisfied. When this occurs, the likelihood is always at a maximum for models with the functional forms in equations (5) to (8). Note that these equations are nonlinear and must be solved using some form of search procedure such as the Newton-Raphson method (Batty, 1976).

Some interpretation is required here for the maximum likelihood methods of estimation. The equations to be met for the count models given in Table 10.1 are straightforward in that they imply that total populations must be replicated by the models and that the relevant mean or logarithmic mean must be met for best estimation. However for the density models, the constraint equation on the totals is a sum of densities while that for the mean is

Table 10.1. Calibration equations for the population density models

	Calibration Method		
	Loglinear Regression	Maximum Likelihood Estimation	
		Scaling Constant	Friction Parameter
Expon Count	$\log P_i = \log K_1 - \lambda_1 r_i$	$\sum_i P_i = P$	$\sum_i P_i r_i = R_1$
Expon Density	$\log \rho_i = \log K_2 - \lambda_2 r_i$	$\sum_i \rho_i = \rho$	$\sum_i \rho_i r_i = R_2$
Power Count	$\log P_i = \log G_1 - \alpha_1 \log r_i$	$\sum_i P_i = P$	$\sum_i P_i \log r_i = R'_1$
Power Density	$\log \rho_i = \log G_2 - \alpha_2 \log r_i$	$\sum_i \rho_i = \rho$	$\sum_i \rho_i \log r_i = R'_2$

an average travel density. These are less meaningful than the count equations although there is no reason a priori why these should not be used. However, there is another set of density estimates which might be derived directly from the count models. Simply by normalizing the predicted counts by area, a set of densities consistent with the counts can be formed. In the same way, another set of counts might be formed by multiplying the predicted densities by the areas to get derived populations. Clearly there are more variants and predictions that might be easily generated from this scheme of combinations and these will be explored in future work.

Integrating diverse software modules

At this point, we will present the ways in which the model variants are meshed with the GIS through the relation of the attribute and digital data to the external operation of the population models. In the following section we will describe how the data base was set up for this involved considerable preprocessing on other machines and with other software before the relevant data could be imported into ARC/INFO. Here we will first concentrate on the methods used to integrate various functions which provides an obvious preliminary to the applications to the Buffalo region illustrated in later sections. The software used here is Version 6 of ARC/INFO, running on the SUNY-Buffalo NCGIA Sun platform, which is composed of some 20 SPARC stations served by a SPARC server. We will now explain how this software was configured into the form relevant to population density modelling in ARC/INFO.

The workstations operate under Sun's own version of X Windows, namely Open Windows which is the medium used to operate the model. In fact, Open

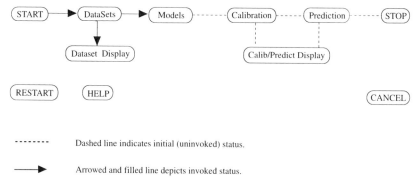

Figure 10.2. The main menu.

Windows is first loaded and this represents the major controlling system although ARC/INFO itself is loaded from SunView which, in turn, is called from Open Windows. The reason for this relates to the fact that some of the introductory graphics were written in Fortran using the CGI graphics library which only operates under SunView. Hence to make use of the graphics, it was necessary to open a shelltool in which the logo is first displayed which in turn opens another shelltool which acts as the basic command window for running ARC/INFO. These two shelltools remain active throughout the processing, and are located in a fixed position at the base of the screen. The shelltool at the bottom left simply displays the logo and title throughout the session while the shelltool which acts as the command window represents the way the user communicates with the program by keyboard, and also displays numeric output from the software. In fact as we shall see the main interactive structure is through the mouse which activates the standard point and click menuing systems available within ARC/INFO.

When these windows have been opened, the user loads ARC/INFO and ArcPlot which in turn enables the main macro to be executed. Apart from the logo and title, the user first sees an introductory screen explaining how interaction with the system is achieved, focusing on the area of application and presenting certain locational characteristics of the region in question giving the user some sense of scale and orientation. The main menu is then displayed at the top of the screen in its centre and this shows the various stages which the user is able to progress through in learning about the data and the model to the point where useful predictions can be made. At present, the introductory screen is being extended to provide a more user-friendly and graphic display of the software's structure and in this, we will be introducing further map data of the region based on remotely sensed and other pictographic data.

The main menu lists the sequence of stages through which the user must pass to operate the software and this is shown in Figure 10.2. In essence, these stages correspond with data exploration, model calibration and prediction

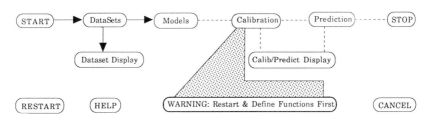

Figure 10.3. Warning for skipping a stage in the flow chart.

(Batty, 1992), and these are identified as involving the selection of *data sets, models* (functional forms), *calibration* (methods) and *prediction* (based on the calibrated models). The user must begin at *start* and end at *stop* making the progression from left to right across the choices in the menu. Once each stage has been completed, the broken line linking the immediately preceding stage becomes a solid arrow line thus indicating where the user is located; if any stage is inadvertently bypassed, the user is warned to *restart* and define the functions which have been missed. This is shown in Figure 10.3. The *cancel* button simply cancels the current choice while the *help* button is designed to provide online help to the user; at present, help is not enabled.

When the user wishes to make a choice, the mouse is pointed at the relevant screen button and the left mouse button is clicked. This will activate a sub-menu which controls the detailed choice of model type to be made; an example is given in Figure 10.4 for the selection of data sets. Figure 10.1 above shows the choices which can be made for the data set in the first stage, model function in the second, and data type, then estimation method in the third calibration stage. These latter choices of data type and estimated model are also executed again in the last prediction stage which in effect simply runs the calibrated models so that graphic predictions can be produced. The choices associated with the first three stages are shown in Figures 10.4, 10.5 and 10.6. The choice from four data sets is straightforward, the choice from three functional forms – exponential, power and gamma is presently restricted to the first two, while the choice of one of two data types in calibration – density or count, precedes the choice of either the loglinear or maximum likelihood method of calibration. The prediction stage is still under construction but it will be similar to the calibration.

Numeric outputs from the models in the form of calibrated parameter values and model performance as well as summaries of predictive data are available when calibration and/or prediction are invoked. The values of the scaling and friction parameters as well as the coefficient of determination (R^2) between observed and predicted population distributions are output in the shelltool ARC/INFO Command Window although the main outputs of the model are from the *display* modules which are used to explore the various data sets and calibrated model predictions. The sub-menu associated with display is shown in Figure 10.7. This consists of three main functions

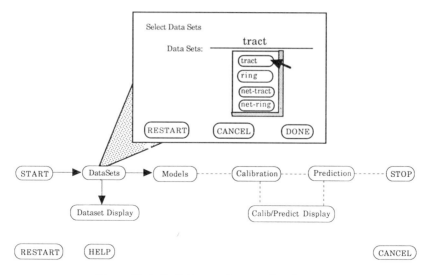

Figure 10.4. Pull-down sub-menu for data sets.

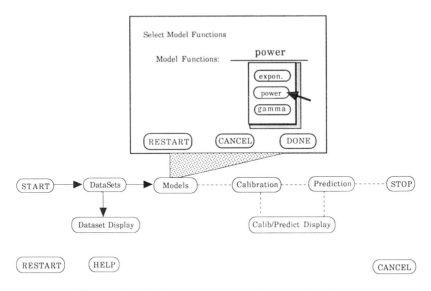

Figure 10.5. Pull-down sub-menu for model functions.

controlled by the *graphics, map,* and *zoom* (*-in* and *-out*) buttons. If graphics
are chosen, three population profiles are graphed against increasing distance
from the CBD. First the counts for each individual tract are shown, followed
by the cumulative counts with increasing distance from the CBD, and then
the individual density values for each tract. These profiles provide an aggregate
sense of how good the population model is likely to be but a more complete

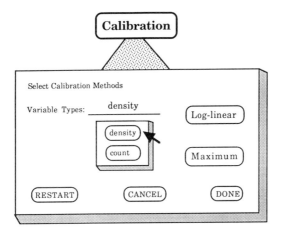

Figure 10.6. Sub-menus for data types and calibration methods.

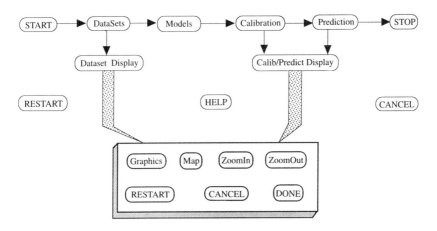

Figure 10.7. The display module.

picture can only be assembled when the data are mapped and this is the function of the *map* button. Two choropleth maps based on counts and densities in each tract with a suitably chosen colour spectrum, are displayed when this map function is invoked, while a map of census tracts is also displayed on the right of the screen so that the user can zoom-in or -out in relation to this basic configuration. When the user decides to zoom-in, the cursor can be used to define the top left and bottom right corners of a window on the map to be enlarged by the zoom. Maps which are zoomed are displayed in the same way as the originals and thus the user can zoom indefinitely to achieve a desired display at any desired scale.

The screen graphics have been laid out in such a way as to communicate spatial patterns as effectively as possible and at present, these are being extended to include three-dimensional surface plots and a variety of other model performance measures. In fact, it is remarkable how flexible the AML language within ARC/INFO is in allowing access to a variety of software external to the main package used. Over 30 separate external program modules are incorporated into this system based on data manipulation, model estimation and prediction programs written in C, a variety of AML and menu modules as well as occasional Fortran programs and screendumps. Two windowing systems – Open Look and SunView – are also used, thus illustrating the variety which can be incorporated into a proprietary GIS. There are some functions which are annoyingly absent and some bugs in the ARC/INFO software with respect to display but in general the system is robust. With respect to the speed of processing, the limits mainly depend on the number of users on the system and the capacity of the network linking the workstation to server rather than the type of code used or even the size of problem. Fortran programs are noticeably slower than C programs although are still within acceptable user response limits. Nevertheless these types of detailed issues are continually under review and improvements to the prototype are being made all the time.

Applications to the Buffalo metropolitan region

In this section, we will describe our application to the Buffalo region through its data, and through its model predictions. First, we need to explain how the data was compiled to the form necessary for the analysis and then we will present a typical 'run' of the prototype so that the reader might get some feel for the capabilities of the software. Although this chapter primarily focuses upon design questions for interfacing GIS with urban models, we will also explain the types of insight into the functioning of the region through exploratory spatial data analysis which the software allows. Finally we will present the results of all possible variants of the models which again, although a by-product of this analysis, is what the more theoretically inspired user is likely to be most interested in.

The development of the database has been comparatively straightforward. We selected seven counties in the hinterland of Buffalo which is the primate city in Western New York (WNY) State. The extent of the region defined by these counties is about 100 kilometres in an east-west direction and about 170 kilometres north-south. The region is bounded by Lake Ontario on the north, by Lake Erie to its south-west, the Niagara River and Peninsula in the west and north-west, with all these features defining the international boundary between the USA and Canada. The foothills of the Alleghenies lie to the southeast and the region is open but rural on its eastern boundary. Greater Buffalo has a population of around 1.049 million and in the seven counties the total population is 1.523 million. The city is very much a border town in that with the Canada–US Free Trade Agreement, there is considerable cross-border activity in all economic sectors and increasingly the local economy is being driven by Toronto some 150 kilometres to the northwest. The ultimate aim of our project is to develop a variety of urban models for the WNY–Southern Ontario region, building on the common border database being developed by the Census Bureau and Statistics Canada. As yet however, our concern is with data exclusively in WNY State.

The seven-county region is composed of some 167 Minor Civil Divisions which are disaggregated into 379 Census tracts. These tracts are the units used for data in the models. The digital boundaries of these tracts were extracted from the post-census version of the TIGER files available on CD ROM while the population data down to block group level was taken from the STF1A Census file which was available for WNY from the State Data Center at Albany. Considerable preprocessing of these files was necessary before the data could be put into the appropriate form. The tapes were first processed on a Vax where considerable time had to be reserved for this purpose while the data was eventually ported to Sun discs prior to its reclassification using ARC/INFO. This process took around two months to complete, much of which was spent sorting and retrieving the correct data; the project was much aided by expertise with the TIGER files which was already available from previous projects (Klosterman and Xie, 1992). Preliminary checks of the data reveal that the counts of population and digital boundaries seem to be appropriately ordered and the simplicity of the data itself has aided this process considerably.

The major problem of application involves the definition of population density and the types of models which are appropriate. Both aspects of the problem led to the definition of four data sets, the form and composition of which we briefly alluded to earlier. With respect to the types of model defined, such models clearly define a two-dimensional density cone around the CBD, of which a one-dimensional cross section constitutes the profile. Strictly speaking, a two-dimensional model should be specified to match this data based on both position (distance from the CBD) and orientation; the models given earlier in equations (1) to (3) were in fact one-dimensional. Such models can be fitted quite consistently to two-dimensional data arranged on the one-dimensional line but if the data were to be aggregated into bands or rings around the CBD, then the appropriate model to fit would be an aggregated version of the two-dimensional equivalent. However such aggregation is appropriate whatever the model selected for it does mask variations in the data set and provides a measure of necessary smoothing.

To summarize then, two types of model based on one- and two-dimensional forms are possible and the data sets to which these models might be fitted should also be in two-dimensional form or aggregated into rings to be treated one-dimensionally. The argument is however a little convoluted because there seems no best data set or model in these terms. However what we have done so far is to fit only one-dimensional models to two-dimensional data collapsed onto a one-dimensional distance profile which we call the *tract* data or aggregated into one-mile rings around the CBD which we refer to as *ring* data. In future applications we will extend these models to two-dimensions and develop additional data sets based on cumulative counts and other aggregations of the data.

A related problem concerns the actual measurement of density. It is obvious that in each Census tract, the entire area is not given over to population. The densities formed from such total measures of area are clearly gross densities and their appropriateness to the 'true' measure of residential density is clearly in doubt. What is required is a measure of net density which is based on a more restricted definition of the area, that associated with residential extent and excluding industrial and large open land. In fact this problem has plagued much population density modelling in the past. However a possible data set for this more restricted definition is available from the TIGER files. In these files, road network features are classified into 8 types of which type 4 relates to neighbourhood streets. It is possible to measure the length of these streets in each Census tract and to use these as a proxy for area. In fact, this measure is more complicated than might seem at first for the higher the density of population, the higher the density of the street network. Better measures might be constructed however from these data based on bounding polygons but this is a matter for further research.

At present two additional data sets based on network length have been calculated and these are referred to as *nettracts* and *netrings*. New counts from these network lengths have been formed for tracts and rings by counting population per mile of street segment length which is clearly a type of density per linear mile rather than a count *per se*. We have computed our density variable as an extension of this count variable in that we have defined an average 'buffer' around the street segment using the appropriate function in ARC/INFO and have measured the areas of the buffers around the street segments in the tract or ring in question. Density is then constructed by dividing the total buffered areas into the raw count. Note that the buffers were taken to be one quarter of the average length of the street segments within the division of Buffalo city. These measures are somewhat controversial as yet and the poor performance of the model results based on these data may be due to these definitions which tend to generate particularly unrealistic measures in tracts on the region's periphery. Considerable refinement of these data is at present underway. Readers should also note that as the nettract and netring counts and densities are in fact two closely related types of density, the results in the later sections for each of these data sets are quite similar. Thus, these four data sets have been used as shown in Figure 10.1 to generate four sets of eight model variants which are reported in the next section.

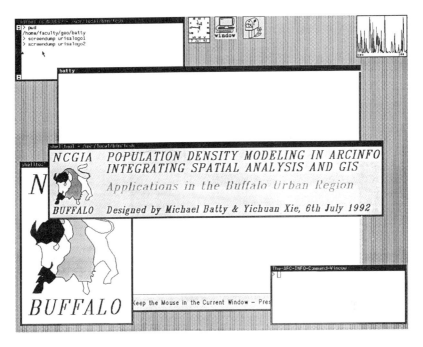

Figure 10.8. The logo in SunView.

Running the software for the Buffalo region

The best way to demonstrate the software to the reader without access to the package is by presenting a visual record of the screen displays generated at each stage. This will provide us not only with a visual trace of the user's progression but also material on which some tentative substantive evaluation of data exploration and model performance can be based. Here we will mainly concentrate on exploring the various data sets used as a basis for modelling although we will show an example of prediction with the best fitting model to give the reader a feel for the extensiveness of possible outputs. Moreover we will concentrate here upon showing three screen pictures from the display module for each of the four data sets: based on the graph profiles, the map distributions and a first zoom-in on these maps. We will present these pictures in their entirety for the tract data set but will cut and paste only the main portions for the other three sets to minimize the number of figures displayed.

In Figure 10.8, we show an enlargement of the logo which is scaled into a SunView window on each run through the software. Although this is based on Fortran code accessing CGI graphics which will only run under SunView, we still have greater control over the graphics in this medium than in ARC/

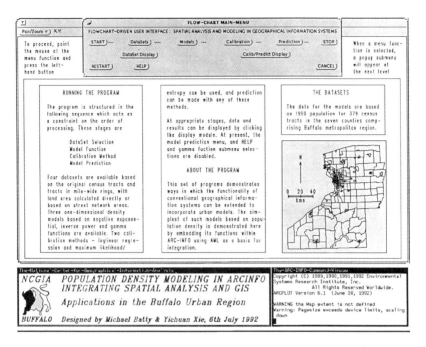

Figure 10.9. The standard screen layout: introduction to the software.

INFO though ArcPlot. The user begins by loading Open Windows which in turn opens the SunView Logo Window in the bottom left of the screen and the active ARC/INFO Command Window at the bottom right from which ARC/INFO is first loaded and run. When this is activated, the ArcPlot canvas fills the rest of the screen, some introduction to the system is displayed including the Census tract map, its scale and orientation, and the main menu as in Figure 10.2 pops up at the top of the screen. At this point the user is looking at the picture shown in Figure 10.9.

The sequence is started by choosing a data set. When one set from the four has been chosen by the appropriate point and click, the user can examine population profiles over distance and maps at scales equal to or finer than the region. We will show these using the example of the tract data set. In Figure 10.10 using the graphics button, we show the typical population profiles which indicate the highly skewed nature of the population–distance relation. The maps of population counts and densities in the Census tracts are shown for the region in Figure 10.11, and for two zoom-ins at successively finer scales in Figures 10.12 and 10.13. We have changed the colours of these maps to reflect proper grey tones but we have used a much narrower range of tones than the actual colours on the screen which range from high density and high counts – purple through blue, to lower values – based on green. In fact these maps do give some sense that the population density is declining with distance

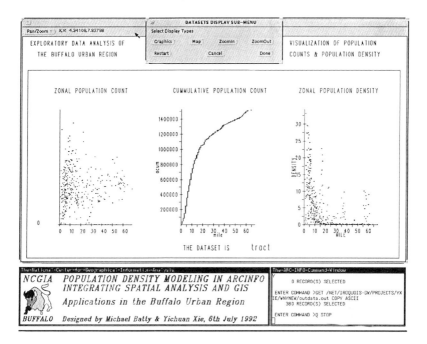

Figure 10.10. Graphics profiles for the tract data set.

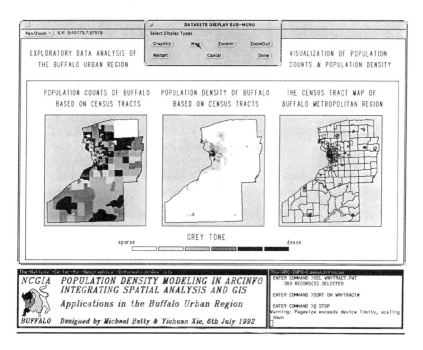

Figure 10.11. Tract map distributions in ArcPlot.

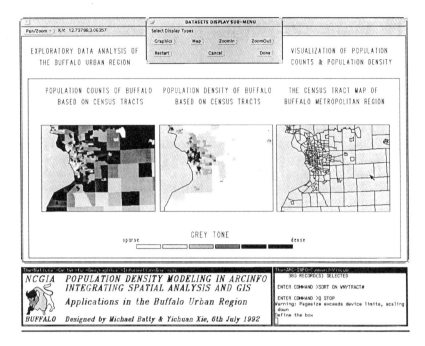

Figure 10.12. An initial zoom-in on map tracts.

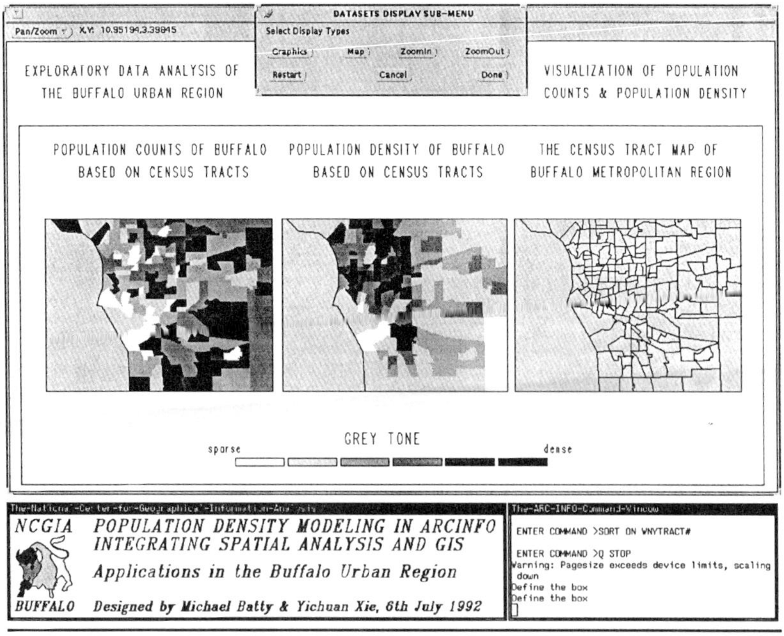

Figure 10.13. A second more detailed zoom.

from central Buffalo but it also reveals that region is too large in that it assumes that towns up to 100 kilometres away from Buffalo are within its density field. This is in fact a problem with our regional definition, which can be detected throughout the data sets.

For the other three data sets, we have stripped off the menus and logos and stitched the profiles, region maps and zoomed maps together into single figures. These are shown in Figures 10.14, 10.15 and 10.16 for the ring, nettract and netring data sets. In fact it is worth commenting upon the ring data sets in Figures 10.14 and 10.16. It is unusual to see such graphic demonstration of these density effects aggregated into rings and this reveals the power of such systems to visualize relationships and to search for new spatial patterns. The last graphic that we will show is taken from the prediction stage of the process. Figure 10.17 shows observed and predicted population densities for the ring data as well as the percentage differences or residuals between observed and predicted values. This type of graphic can also be produced for counts, and it can be generated for counts derived from predicted densities and densities derived from predicted counts. In short for each data set, four such graphics are possible, thus showing how the system can explode in its ability to handle different model and data options, something which is fairly novel in most areas of modelling where the choice of model has not been formally and explicitly possible hitherto.

Before we turn to the model results for these four data sets, it is worth noting significant points immediately revealed through these graphics. In Figure 10.10, the scatter of data points for the counts does in general reveal a mild fall-off in volumes of population as distance from Buffalo increases although zones get larger in area to compensate. However the fall-off is confirmed in the cumulative population counts and apart from some obvious outliers becomes even clearer in the density profiles. In Figures 10.11 to 10.13, the deadheart of Buffalo itself is clearly revealed as are the generally higher densities in the urban area itself. The deadheart of the city is even clearer when the ring data is examined in Figure 10.14. In general the ring data is the best set for revealing the monotonic decline of densities with distance from the CBD and this is clearly seen in the density profile in the top diagram in Figure 10.14. The nettract data in Figure 10.15 seems little different from the tract data although the netring data in Figure 10.16 is counter-intuitive in that several densities in the most peripheral zones of the region are quite clearly out-of-line. Finally, Figure 10.17, which shows the predictions of the ring model, is simply illustrative of what we are doing at this stage of the process and has no substantive interpretation as yet.

Finally, we should say something about model performance. For each of the 32 models shown in Figure 10.1, we have distance parameter estimates and measures of association between observed and predicted dependent variables based on R^2. We will not show the scaling parameter as these do not have the same substantive interpretation as the friction of distance parameter values. Moreover, we have arranged these results so that the

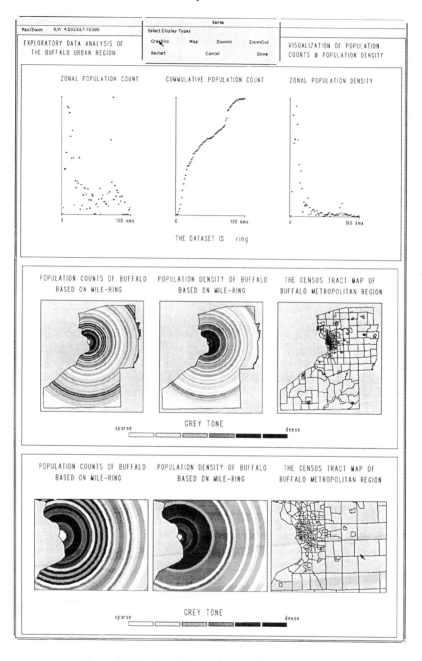

Figure 10.14. Graphics and maps for the ring data.

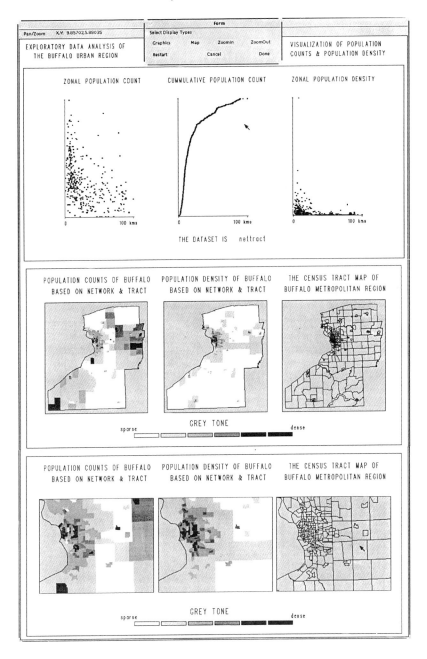

Figure 10.15. Graphics and maps for the nettract data.

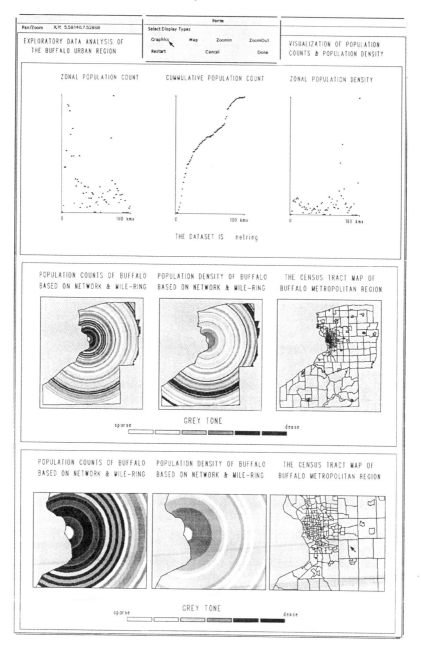

Figure 10.16. Graphics and maps for the netring data.

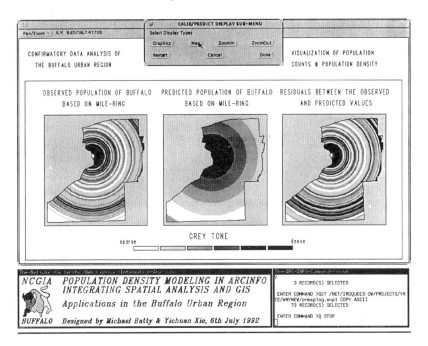

Figure 10.17. Predictions in map form for a ring model.

comparisons between functional forms are grouped first, then density and counts, methods of calibration, and finally data sets. This grouping enables parameters with the same order of magnitude values to be directly compared. These values are shown in Table 10.2 where it is immediately clear that the overall performance of these models is poor. The best are those based on the ring data and it is from these models that the predictions shown in Figure 10.17 have been generated. What however these results do show is the need for much better spatial definition. The density models should not be applied to the entire region as we have done because it is clear that towns in the periphery with their own focus must be excluded. Their observations look like outliers as well they might be in terms of the Buffalo urban field. Moreover, it is clearly necessary to firm up on the types of functions (one- or two-dimensional) required and upon the ways in which density is measured. There is much to do on the network measures of area, yet this innovation is still promising.

Conclusions: the ongoing research program

As the work reported here is part of an ongoing research program, there are many improvements and extensions underway and we will note some of these

Table 10.2. Model parameters and performance

Model	Data type	Calibration Method	Data Set	Parameter α	R^2
expon	density	loglin	tract	−0.0609	0.2912
			ring	−0.0604	0.6374
			nettract	−0.0179	0.1570
			netring	0.0151	0.0794
		maxl	tract	−0.0439	0.2099
			ring	−0.0950	0.9979
			nettract	−0.0173	0.1513
			netring	0.0275	0.1441
	count	loglin	tract	0.0012	0.0006
			ring	−0.0185	0.1154
			nettract	−0.0180	0.1546
			netring	0.0152	0.0771
		maxl	tract	0.0007	0.0003
			ring	−0.0222	0.1388
			nettract	−0.0174	0.1489
			netring	0.0276	0.1399
power	density	loglin	tract	−1.1222	0.3387
			ring	−1.4062	0.7243
			nettract	−0.3353	0.1877
			netring	0.1492	0.0161
		maxl	tract	−2.9544	0.8916
			ring	−3.3092	0.5867
			nettract	−2.7006	0.6616
			netring	−2.2486	0.2431
	count	loglin	tract	0.0372	0.0018
			ring	−0.4325	0.1326
			nettract	−0.3364	0.1852
			netring	0.1478	0.0153
		maxl	tract	−2.4727	0.1254
			ring	−2.5145	0.7709
			nettract	−2.7010	0.6724
			netring	−2.2485	0.2334

here as conclusions. First, with respect to the application, it is clear that the spatial definition of regional extent is too great and that a more restricted focus on the immediate urban field of Buffalo needs to be decided. This has now been added and is under user control as part of a module enabling the user to define the level of aggregation from the data. In a complementary way but for different purposes, models need to be run at different levels of spatial aggregation so that their comparative performance may be evaluated in the search for meaningful levels and in the quest to explore the modifiable areal unit problem. In short, the software is being extended to deal with the whole question of spatial aggregation in an exploratory as well as confirmatory context.

With respect to the data sets, other more refined measures of density based on network length need to be developed. A further data type based on cumulative counts of population needs to be used in fitting the models; this variable has already been presented as part of the graphics unit of the display module and is a type which represents another portrayal of density and count. We have already mentioned that the suite is being extended to deal with the gamma function in terms of model form, and extending the models to two-dimensional from one-dimensional form is also on the agenda. It is likely too that we will add another method of estimation based on modified maximum likelihood while there are possibilities for developing fractal methods of parameter estimation for the power functions (see Batty and Kim, 1992). If we assume that the software is extended to deal with 3 levels of spatial aggregation (say, at block group, Census tract and MCD), that there are 6 rather than 4 data sets, 3 definitions of regional extent, 3 data types, 3 functional forms, 2 dimensional specifications and 3 methods of estimation, this generates 2,916 possible models to test, thus illustrating the value of the framework in not only implementing such models on large data bases where GIS is used as the user interface but also enabling large numbers of model forms based on different data sets to be thoroughly explored. This is an approach which has hardly been possible hitherto and is a direct consequence of having these powerful GIS display engines now available.

There are other longer term extensions in mind as well. First, extending the data base to Southern Ontario is an important priority and this will bring a completely new set of research problems relating to data integration as well as cross-border substantive differences in applicability of these models. Extending the density models to different categories of population and thence to other activities such as employment is an obvious possibility with some basis too for comparison between different activities in terms of density profiles. This chapter is very much a report on work in progress and we fully expect to be able to report the extensions noted above in the near future. The software in its present form is very much a prototype although it is, in our view, a sufficiently realistic demonstration of the flexibility of current GISs, such as ARC/INFO, in their ability to integrate spatial statistics and predictive models.

Acknowledgement

This chapter represents work that is related to research being carried out at the National Center for Geographic Information and Analysis, supported by a grant from the National Science Foundation (SES-88-10917). The Government has certain rights in this material. Support by NSF is gratefully acknowledged. Any opinions, findings and conclusions or recommendations expressed in this material are those of the authors and do not necessarily reflect the views of the NSF.

References

Alonso, W., 1964, *Location and Land Use*, Cambridge, Mass.: Harvard University Press.
Anas, A., 1987, *Modelling in Urban and Regional Economics*, New York: Harwood.
Angel, S. and Hyman, G. M., 1976, *Urban Fields: A Geometry of Movement for Regional Science*, London: Pion Press.
Batty, M., 1974, Urban density and entropy functions, *Journal of Cybernetics*, **4**, 41–55.
Batty, M., 1976, *Urban Modelling: Algorithms, Calibrations, Predictions*, Cambridge: Cambridge University Press.
Batty, M., 1978, Urban Models in the Planning Process, in D. T. Herbert and R. J. Johnston (Eds.), *Geography and the Urban Environment: Volume 1: Progress in Research and Applications*, Chichester: John Wiley, pp. 63–134.
Batty, M., 1992, Urban models in computer graphic and geographic information system environments, *Environment and Planning B*, **19**, 663–685.
Batty, M. and Kim, K. S., 1992, Form follows function: reformulating urban population density functions, *Urban Studies*, **29**, 1043–1070.
Beckmann, M. J., 1969, On the distribution of urban rent and residential density, *Journal of Economic Theory*, **1**, 60–67.
Bell, G., 1992, Ultracomputers: a teraflop before its time, *Science*, **256**, 64.
Bussiere, R., 1972, Static and dynamic characteristics of the negative exponential model of urban population distributions, *London Papers in Regional Science*, **3**, 83–113.
Clark, C., 1951, Urban population densities, *Journal of the Royal Statistical Society*, A, **114**, 490–496.
Densham, P., 1991, Spatial Decision Support Systems, D. J. Maguire, M. F. Goodchild, and D. W. Rhind (Eds.), *Geographical Information Systems: Principles and Applications*, London: Longman, pp. 403–412.
Densham, P., 1993, Integrating GIS and Spatial Modelling: The Role of Visual Interactive Modelling in Location Selection, *Geographical Systems*, **1**, in press.
Ding, Y. and Fotheringham, A. S., 1992, The integration of spatial analysis and GIS, *Computers, Environment and Urban Systems*, **16**, 3–19.
Gatrell, A. and Rowlingson, B., 1993, Spatial Point Process Modelling in a GIS Environment, Chapter 8, this volume.
Harris, B., 1989, Beyond Geographic Information Systems: computers and the planning professional, *Journal of the American Planning Association*, **55**, 85–90.
Harris, B. and Batty, M., 1993, Locational models, geographic information and planning support systems, *Journal of Planning Education and Research*, **12**, 102–15.
Huxhold, W. E., 1991, *An Introduction to Urban Geographic Information Systems*, New York: Oxford University Press.
Kehris, E., 1990, A Geographical Modelling Environment Built Around ARC/INFO, Research Report 16, North West Regional Research Laboratory, University of Lancaster, Lancaster, UK.
Klosterman, R. E., Brail, R. K. and Bossard, E. G. (Eds.), 1993, *Spreadsheet Models for Urban and Regional Analysis*, New Brunswick, NJ. Center for Urban Policy Research, Rutgers University.

Klosterman, R. E. and Xie, Y., 1992, The TIGER family: What's available, *Microsoftware News For Local Governments*, **9**, 1–5.

March, L., 1971, Urban systems: a generalized distribution function, *London Papers in Regional Science*, **2**, 156–170.

Newton, P. W., Taylor, M. A. P. and Sharpe, R. (Eds.), 1988, *Desktop Planning: Advanced Microcomputer Applications for Physical and Social Infrastructure Planning*, Melbourne, Australia: Hargreen Publishing Company.

Openshaw, S., 1990, Spatial Analysis and Geographical Information Systems: A Review of Progress and Possibilities, in H. J. Scholten and J. C. H. Stillwell (Eds.), *Geographical Information Systems for Urban and Regional Planning*, Dordrecht and Boston, Mass.: Kluwer Academic Publishers, pp. 153–163.

Rogerson, P. A. and Fotheringham, A. S., 1993, GIS and Spatial Analysis: Introduction and Overview, Chapter 1, this volume.

Wilson, A. G., 1970, *Entropy in Urban and Regional Modelling*, London: Pion Press.

Wise, S. M. and Haining, R. P., 1991, The Role of Spatial Analysis in Geographical Information Systems, *Proceedings of the AGI'91 London*, Westrade Fairs, London, pp. 3.24.1–3.24.8.

11

Optimization modelling in a GIS framework: the problem of political redistricting

William Macmillan and Todd Pierce

Introduction

The quantitative revolution in geography and related disciplines led to the development of a wide range of mathematical models which promised rather more than they delivered. Much the same might be said of the computing revolution. In the former case, part of the problem was the absence of adequate data and appropriate software. In the latter case, the difficulty arguably centres on the limited mathematical modelling capabilities of contemporary Geographical Information Systems. Thus, there seems to be a good *prima facie* case for expanding GIS modelling facilities and for building or rebuilding models in a GIS framework. Such developments should be mutually beneficial and reinforcing.

Over the past couple of decades, one of the major advances in geographical modelling has revolved around mathematical programming or optimization. In particular, urban land use and transportation systems, which have been studied in a variety of ways over the years, are now capable of being analysed in a highly sophisticated fashion with the aid of optimization models (see Wilson *et al.*, 1981). As well as this descriptive use – describing the ways sets of economic agents behave in spatial economies – optimization models are also used prescriptively. That is, they are used as vehicles for expressing decision making problems and for exploring their solutions. It follows that a GIS with optimisation modelling capabilities would have the potential both to analyse a wide range of geographical systems and to operationalize the planning problems associated with such systems. Indeed, a spatial decision support system worthy of the name should certainly be capable of casting decision problems into an optimization form.

This chapter explores one of a number of related attempts to introduce optimization models into a GIS framework. The particular issue considered here is that of congressional redistricting in the USA. In this context, optimization is used prescriptively or, in other words, in a decision support role. Other work in hand is concerned with the use of optimization models in spatial economic analysis and in environmental–economic project evaluation with particular reference to big dam projects.

The problem of congressional redistricting has a number of interesting features mathematically. It is also a problem which has attracted the attention of GIS producers. To the best of our knowledge, the only redistricting systems currently available are passive ones in which a user can select individual areas to be moved between electoral districts. These systems provide a piecemeal approach to the problem which leaves plenty of scope for subtle gerrymandering (which may, of course, explain their popularity with legislatures). In mathematical terms, passive systems are extremely simple. On the other hand, there are a large number of models or algorithms, which predate the GIS era, that were designed to produce acceptable or optimal districting plans with respect to clearly defined criteria. All of these algorithms are unsatisfactory in one way or another. What is not available, again to the best of our knowledge, is a GIS system containing an active and, therefore, mathematically sophisticated redistricting algorithm. This, then, is one of those areas where GIS functionality could be improved and, hopefully, models enhanced, by some judicious cross-fertilization.

One of the GIS systems that can be used for passive redistricting is TransCAD. As it happens, this system also contains a number of built-in optimization routines. These include modules for solving the shortest path, travelling salesman, entropy maximization, and facility location problems. Each module is a free-standing 'procedure'; there are no generic optimization routines underlying the separate modules. Of greater significance is the fact that TransCAD is capable of having user-produced algorithms grafted onto it as extra procedures. It should be possible in principle, therefore, to embed a mathematical programming toolkit in TransCAD. The utility of such a toolkit would depend, of course, on the extent to which the empirical problems to be solved can be cast into standard mathematical programming model forms. There are many such forms and a variety of algorithms exists for their solution, most of which are available in commercially produced packages. The Numerical Algorithms Group (NAG) library is the major collection of UK optimization routines. NAG's E04 and H chapters contain unconstrained, linear, integer linear, quadratic, and general non-linear programming algorithms. Although this set is by no means exhaustive, the general task of embedding a mathematical programming toolkit into a GIS framework would certainly be facilitated if these NAG products could be used 'off-the-peg'.

This chapter describes part of the work involved in the development of a set of active redistricting algorithms based on mathematical optimization models and their operationalization as TransCAD procedures. Some of these algorithms utilise NAG routines but it is argued that the mathematically interesting features of the redistricting problem make it desirable to use non-standard algorithms. In particular, a simulated annealing approach is described which appears to be successful in circumventing the difficulties inherent in the structure of the redistricting problem.

Section 2 looks at the mathematics of redistricting and at the possibility of casting the redistricting problem into a standard mathematical form. It is argued that the need to impose a contiguity restriction on districts imposes constraints which make the feasible region of the programming problem non-convex and it is this that leads to the choice of a simulated annealing algorithm. Section 3 describes simulated annealing, whilst Section 4 considers its application to redistricting using an algorithm called ANNEAL. Section 5 gives details of TransCAD-ANNEAL, a prototype redistricting decision support system (DSS). The final section explores alternatives to the approach embodied in ANNEAL and considers possible improvements to the redistricting DSS.

The mathematics of the redistricting problem

After every census, each state in the USA has to draw up new electoral districts. Malapportionment is illegal so at least one of the redistricting criteria has to contain some notion of population equalization. Indeed, this may be the only criterion. In practice, it is generally assumed that districts must also be contiguous (i.e. that no county or group of counties may be isolated from the other counties in its district) although the recently published plans for North Carolina show that contiguity is sometimes neglected (*The Economist*, 1992). Other criteria may be introduced as will be shown below. Finally, there are two technical constraints: that each district must contain at least one county; and that each county must belong to one and only one district. This last constraint is a logical one and is not intended to prevent district boundaries cutting county lines: if it is potentially desirable to split counties between districts, then counties should be disaggregated into wards; the constraint would then be that each ward must belong to one and only one district, thereby preventing it from being wholly in District A as well as being wholly in District B.

Feasible plans and a redistricting objective

It is possible to describe a plan for arranging C counties into D districts by a vector

$$\mathbf{x} \equiv (x_1^1, \ldots, x_C^1; \ldots; x_1^D, \ldots, x_C^D)$$

where, for $i = 1, \ldots, C$ and $d = 1, \ldots, D$

$$(x_i^d = 1) \Rightarrow (\text{county } i \text{ is in district } d);$$
$$(x_i^d = 0) \Rightarrow (\text{county } i \text{ is not in district } d).$$

A district plan can then be said to be feasible if it satisfies the conditions

$$\sum_i x_i^d \geq 1 \qquad \forall d \tag{1}$$

$$\sum_{d} x_i^d = 1 \qquad \forall i \tag{2}$$

$$x_i^d = 0 \text{ or } 1 \qquad \forall id \tag{3}$$

Condition (1) is that every district contains at least one county whilst conditions (2) and (3) together ensure that every county is in one and only one district. The population equality criterion can be formalized as a requirement to

$$\text{minimize } \sum_{d} \left(\sum_{i} p_i x_i^d - P/D \right)^2 \tag{4}$$

where p_i is the electoral population of county i and P is the total electoral population of the state. That is, the criterion may be interpreted as a requirement to minimize the sum of the squares of the deviations of the district populations from their target population, where the target in each case is to have exactly P/D of the state's P electors. Objective (4) together with constraints (1), (2) and (3) constitutes a quadratic integer programming problem. This will be referred to below as *Problem 1*.

Existing redistricting algorithms

The integer nature of this problem helps to explain the variety of approaches that have been adapted to the production of redistricting plans. It leads to the representation of the problem's solution space as a tree. Figure 11.1 depicts the tree for a problem with four counties which have to be allocated to two districts. The bottom line in the figure contains all possible allocations. Whether or not an allocation will be contiguous will depend on the county map. Two maps are shown in the figure. The contiguous allocations associated with each map are labelled 'C' and the non-contiguous ones labelled 'N'. The question of contiguity will be returned to below.

Two popular algorithmic strategies are illustrated in Figure 11.1. One is based on a seeding procedure in which a seed county is selected for each district and the districts are then grown by adding counties to the seeds until they have all been assigned. This strategy is represented in the figure by the selection of the seed (1,2), i.e. by the initial allocation of County 1 to District 1 and County 2 to District 2. All possible assignments of the remaining counties are represented by the boxes on the bottom row of the figure that are connected to (1,2) from below (see the thickened lines). The growth process guides the solution procedure down through the network (down one path leading from the seed) but not in an optimal fashion. The strategy evidently suffers from the fact that it searches only a part of the solution space so there is certainly no guarantee that a global minimum population deviation will be found. Moreover, the rules for the growth process make it highly unlikely that even a local optimum will be reached. The virtue of the strategy is that

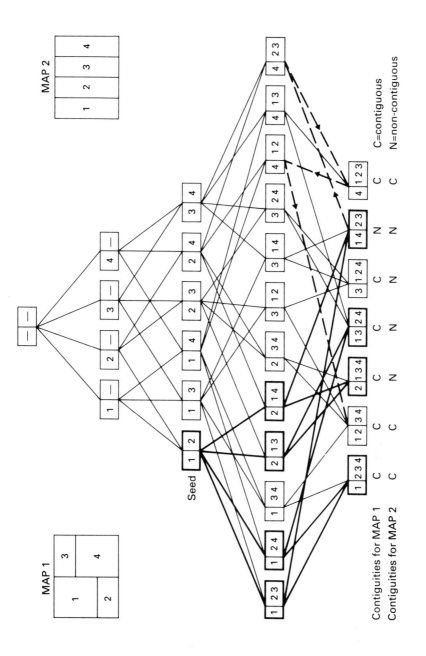

Figure 11.1. The solution tree for a simple redistricting problem.

it does at least deliver feasible solutions. Growth algorithms have been produced by Vickery (1969), Gearhart and Liittschwager (1969), and Rossiter and Johnston (1981).

Table 11.1. Example populations for Map 1 in Figure 11.1

District	Population
1	6
2	5
3	4
4	5

The second strategy involves explicit optimization and swapping. Starting from a complete plan (i.e. from a box on the bottom row of Figure 11.1), a county is swapped from one district to another so that the mean squared deviation is reduced at the fastest possible rate. Swapping continues until the 'best' plan is found (see the dashed lines starting from plan (1 4, 2 3)). This strategy also has its limitations. For example, if the counties are connected as in Map 1 and the populations of the districts are as shown in Table 11.1, then the plan (1 2, 3 4) would be suboptimal but would inhibit further swaps since any single swap (i.e. the detachment of a county from one district and its attachment to another) would lead to a deterioration in the mean squared population deviation. The swapping algorithm produced by Kaiser and Nagel is described in Nagel (1965).

Another optimization strategy is due to Weaver and Hess. Their algorithm creates districts by maximizing district compactness measured in terms of population distribution (Weaver and Hess, 1963). This idea will be taken up in a modified form at the end of this chapter.

The contiguity problem

In addition to the difficulties referred to above, the desire to produce contiguous plans creates a formidable problem. Informally, a contiguous plan is one in which each county in a district is connected to every other county via counties that are also in the district. Put another way, a plan is contiguous if no district has island counties or groups of counties separated from the main body of the district. This idea can be formalized as a contiguity restriction involving powers of the problem's connectivity matrix:

A connectivity matrix K has elements k_{ij} which are 1 if i and j are simply connected (i.e. touching) counties and are zero otherwise.

The connectivity matrix for a district d consists of elements $k_{ij}x_i^d x_j^d$. Thus, if counties i and j are both in d and are touching, the element ij of the district connectivity matrix will be 1. Otherwise it will be 0. To ensure that the counties of an N-county district are contiguous it is sufficient to check that the $(N-1)$th power of this matrix has no zero terms.

If

$$k_{ij}^{d1} \equiv k_{ij} x_i^d x_j^d$$

$$k_{ij}^{d2} \equiv \sum_n k_{in}^{d1} k_{nj}^{d1}$$

$$k_{ij}^{d3} \equiv \sum_n k_{in}^{d2} k_{nj}^{d1}$$

.
.
.

$$k_{ij}^{d(N-1)} \equiv \sum_n k_{in}^{d(N-2)} k_{nj}^{d1}$$

then the contiguity condition takes the form

$$k_{ij}^{d(N-1)} \geq 1 \qquad \forall ij \text{ such that } x_i^d x_j^d = 1.$$

Alternatively, since $x_i^d x_j^d = 1$ only if both i and j are in d, the condition may be written

$$k_{ij}^{d(N-1)} - x_i^d x_j^d \geq 0 \qquad \forall ij$$

Assuming that there are C counties in all, the maximum number there could be in any one district is $C - D + 1$. It follows that a general contiguity condition that might be appended to conditions (1), (2), and (3) is

$$k_{ij}^{d(C-D+1)} - x_i^d x_j^d \geq 0 \qquad \forall ijd$$

where $k_{ij}^{d(C-D+1)}$ is defined recursively as above.

To cope with the contiguity requirement, then, the rather simple optimization problem defined by expressions (1) to (4) inclusive (*Problem 1*) takes the following much nastier form.

Problem 2

$$\text{minimize} \sum_d \left(\sum_i p_i x_i^d - P/D \right)^2$$

s.t.

$$\sum_i x_i^d \geq 1 \qquad \forall d$$

$$\sum_d x_i^d = 1 \qquad \forall i$$

$$k_{ij}^{d1} = k_{ij} x_i^d x_j^d \qquad\qquad \forall ijd$$

$$k_{ij}^{d2} = \sum_n k_{in}^{d1} k_{nj}^{d1} \qquad\qquad \forall ijd$$

$$k_{ij}^{d3} = \sum_n k_{in}^{d2} k_{nj}^{d1} \qquad\qquad \forall ijd$$

.
.
.

$$k_{ij}^{d(C-D+1)} = \sum_n k_{in}^{d(C-D)} k_{nj}^{d1} \qquad \forall ijd$$

$$x_i^d = 0 \text{ or } 1 \qquad\qquad \forall id$$

The nastiness of this problem springs partly from its size, both in terms of its dimensionality and number of constraints, but mainly from the fact that the space defined by the constraints is non-convex. Whereas the feasible region in *Problem 1* would be convex without the integer restriction (constraint (3)), the removal of this restriction from *Problem 2* still leaves a non-convex region in which the objective function is likely to reach numerous non-global, local optima. It is this characteristic of the feasible region that makes solving the redistricting problem very difficult. An approximation to the solution of *Problem 1* could be found by removing constraint (3) and applying standard quadratic programming routines. Indeed, NAG has a prototype integer quadratic programming routine which can be applied to Problem 1 directly. With *Problem 2*, stripping out constraint (3) does not yield a standard model form to which a standard algorithm can be applied. Hence the need exists to use a novel procedure like simulated annealing.

Contiguity checks

The nature and application of simulated annealing will be described below. First, a little more needs to be said about the contiguity constraints since any acceptable redistricting algorithm needs to be able to deliver contiguous plans.

First note that the contiguity check contained in *Problem 2* is over-specified in the sense that all but one of the districts on all iterations will, necessarily, have less than the maximum $(C - D + 1)$ counties. The highest power to which the district connectivity matrix needs to be raised in practice is $N - 1$, where N is the number of counties in the district. Indeed, a smaller power might well suffice since district contiguity is established as soon as all the terms in the Mth. power matrix are non-zero, where M is any positive integer less than or equal to N. It follows that some degree of computational efficiency could be achieved by relegating the contiguity check to a constraint on the operation of the algorithm. That is, a check could be made on each iteration to ensure that the new plan is contiguous. This does not, of course, get round the problem of the existence of non-global local optima but it does provide a way of streamlining any search procedure for the global optimum.

In fact, it is possible to do much better than this, once it has been decided to relegate the contiguity check to a rule in the algorithm. If a feasible and contiguous starting solution is used, it is possible to ensure that the search procedure will be restricted to feasible and contiguous solutions by a simple contiguity check based on a topological observation. It is easiest to explain this observation by using an example.

For the 5-county, 2-district problem illustrated in Figure 11.2, suppose that county 4 has been selected to leave the shaded district (district *d*). It can be seen that county 4 can leave the district without loss of contiguity because the boundary of 4 has only two sections: an interior section along which 4 is joined to other counties in the district; and an exterior section along which there are counties belonging to other districts (including county/district 0 which represents the rest of the world). As the boundary has only two sections, there are only two points along it at which there is a switch in the type of boundary (interior to exterior or vice versa).

Starting from point A on the boundary and proceeding anticlockwise, it is possible to count the number of switching points. Generally, a junction on the boundary is a switching point if the district membership of the two counties that meet at the junction changes from 'the same as the object county' to 'not the same as the object county' or vice versa. Thus, the first switching point is point B because county 3 is in the same district as county 4 but county 0 is not in the same district as 4. The second switching point is point A. Algebraically, the number of switching points is given by

$$(x_1^d - x_3^d)^2 + (x_3^d - x_0^d)^2 + (x_0^d - x_5^d)^2 + (x_5^d - x_2^d)^2 + (x_2^d - x_1^d)^2$$

so the contiguity check is that county 4 can be detached from the shaded district provided

$$(x_1^d - x_3^d)^2 + (x_3^d - x_0^d)^2 + (x_0^d - x_5^d)^2 + (x_5^d - x_2^d)^2 + (x_2^d - x_1^d)^2 = 2$$

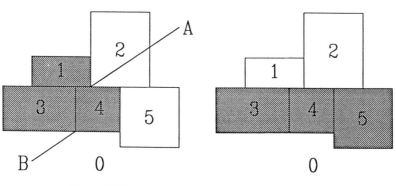

Figure 11.2.   *Figure 11.3.*

It is easy to check that the number of switching points in Figure 11.3 is 4 so county 4 cannot be detached from the shaded district shown there without loss of contiguity. For a county entirely surrounded by counties of the same district, the number of switching points will be zero so the contiguity check would prevent removal again.

One minor complication arises because of the possibility illustrated in Figure 11.4. Here, the cross-hatched county could be detached from the single-hatched district without loss of contiguity, even though it has 4 switching points on its boundary. This problem can be avoided by starting the search procedure from a situation in which there are no island districts (like the one in the centre of Figure 11.4) and ensuring that none are created by disallowing the addition of a county to a district if the county would have 4 switching points when added to the new district (if the cross-hatched county was not a member of the single-hatched district, it is clear from Figure 11.4 that it would acquire 4 switching points if it was to be added to that district).

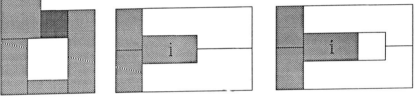

Figure 11.4.   *Figure 11.5.*   *Figure 11.6.*

The form of the topological restriction will, of course, vary with the topology of the system of counties. It may be possible to formulate a general statement of the restriction but we have not produced one yet which covers all cases. For example, the restriction

$$\sum_j \sum_h c_{jh}(c_{ij}x_i^d x_j^d - c_{ih}x_i^d x_h^d)^2 = 4$$

is adequate for cases like that illustrated in Figure 11.5 but not for cases like that in Figure 11.6. The implication of having no general formula is that a set of contiguity equations has to be input by the user for each state for which a district plan is required.

Treating the contiguity restriction as an algorithmic rule enables us to revert to the use of *Problem 1*. However, in doing so, it must not be forgotten that the (non-integer) nonlinearity is still present implicitly.

2.5. Other redistricting criteria

In addition to the population equality and contiguity criteria, numerous other criteria might be employed, including the following:

- high minority representation
- preservation of communities of interest
- partisan balance
- incumbent protection
- high geographic or economic contiguity.

Each of these notions can be quantified and converted into a constraint or an objective to be grafted onto *Problem 1* (Pierce and Macmillan, 1994). Treating an additional criterion as an objective, the objective function of *Problem 1* could be transformed into the weighted sum of two (or more) objectives. Suppose, for example, that the criterion of population equality was to be mixed with that of achieving high minority representation. One way of realising the latter goal would be to concentrate minorities in a small number of districts and this could be done by maximizing the deviation of district minority populations from the average minority population. That is, the second criterion could be

$$\text{maximize} \quad \sum_d \left(\sum_i m_i x_i^d - M/D \right)^2$$

where m_i is the (or a) minority population of county i and M is the total population of the minority over all counties. The objective function of *Problem 1* could then be written

$$\text{minimize} \left\{ w_1 \sum_d \left(\sum_i p_i x_i^d - P/D \right)^2 - w_2 \sum_d \left(\sum_i m_i x_i^d - M/D \right)^2 \right\}$$

where w_1 and w_2 are positive weights such that $w_1 + w_2 = 1$ (minimizing a negative quantity is, of course, the same as maximizing its positive). Any number of criteria could be dealt with in this way. The alternative is to treat some or all of the additional criteria as constraints. This involves choosing a level for each criterion that must be achieved or improved upon. One way of deciding upon the level is to solve the redistricting problem using the criterion in question as the sole objective then taking a percentage of the optimal value as the required level. Thus, for example, the above additional criterion could be reformulated as the constraint

$$\sum_d \left(\sum_i m_i x_i^d - M/D \right)^2 \geq w_3 N^*$$

where w_3 is a weighting which is less than 1 and N^* is the optimal value of

$$\sum_d \left(\sum_i m_i x_i^d - M/D \right)^2 \quad \text{when it is maximized subject only to the constraints}$$

in *Problem 1* (with or without the contiguity restriction).

The approach we have adopted for dealing with multiple criteria is to treat them all as fuzzy. In fuzzy optimization, no distinction is made between objectives and constraints. Each can be formulated as a technically equivalent goal or restriction and a single criterion results. A discussion of fuzzy optimization and details of its use in the present context are given in Pierce and Macmillan (1994).

Simulated annealing

We now turn to simulated annealing, which was suggested as a strategy for solving the redistricting problem by Michelle Browdy (Browdy, 1990). For Browdy, any computer redistricting algorithm must yield only one solution irrespective of initial conditions; furthermore, this unique solution should be a global optimum of an objective function. Simulated annealing, an optimization procedure of recent vintage and considerable promise, satisfies these requirements. The process uses random elements to overcome dependence on initial conditions, thereby bringing the global optimum within reach (Browdy, 1990). Several large optimization problems, considered unsolvable when using traditional methods have been solved by simulated annealing.

A basic overview of simulated annealing is found in Press *et al.* (1986). The procedure is modelled on the physical process of annealing. When a metal or crystal in liquid form is cooled slowly, with the system allowed to reach thermal equilibrium after each temperature drop, atoms are able to lose energy at a rate slow enough to permit the creation of an ordered lattice of atoms. At each temperature, the system can be at any energy level. The probability of the system being at a given energy level is a function of $\exp(-E/T)$, where E is the energy level and T is the system temperature. Therefore, as the temperature is lowered, the system always has a chance of changing to a higher energy state. This probability lessens, though, as the temperature is lowered. Eventually, when the system approaches zero temperature, the atoms approach their minimum energy state. In contrast, quick cooling or quenching results in a higher energy state. In quenching, the system is not allowed to reach thermal equilibrium after each temperature change. With less time being available for atoms to reach an ordered atomic arrangement, the atoms lock into an irregular structure.

Traditional optimization algorithms can be compared to quenching; in the rush to find a solution, only local optima may be reached (Press *et al.*, 1986). The true global optimum, corresponding to the minimum energy state, is generally not obtained. Just as in quenching, no changes to a higher energy level are allowed, and the algorithm becomes trapped in the vicinity of a local minimum. Simulated annealing, though, solves an optimization problem slowly until the minimum state is reached. The energy level for any state of the system has to be defined and a probability function is required to determine the chances of the system going from a given energy level to a higher or a lower one. Because there is always a chance of the system jumping from a given energy state to a higher one, the algorithm avoids (or should avoid) becoming trapped at or near a local minimum. By slowly lowering the system's temperature, the energy state is lowered as well, although occasional energy rises are needed to escape local minima. As the energy level is reduced, the chances for rises in the level are also reduced. Eventually, the energy can be lowered no more. The system has approached the global minimum.

Simulated annealing had its start in the algorithms of Metropolis *et al.* (1953), developed to model the behaviour of systems of particles such as a gas in the presence of an external heat bath (Johnson *et al.*, 1989). The system starts in a given state with a certain energy level. If the system is given a small random perturbation, the resulting state has a different energy. The difference in energy levels, ΔE, determines whether the new state is accepted or not. If ΔE is negative, indicating a decrease in energy level, then the change is accepted. If ΔE is positive, indicating an increase in energy level, the change is accepted with probability $\exp(-\Delta E/T)$.

The Metropolis algorithm was combined with an annealing schedule, or decreasing agenda of temperatures, by Kirkpatrick *et al.* (1983) and independently by Cerny (1985) to create the optimization procedure of simulated annealing. Applying simulated annealing requires interpreting every possible solution as a state of the system. The value of the objective function for each solution becomes the energy level. A control parameter, still called temperature, is gradually lowered. At each value of temperature, possible changes to the current solution are identified and are implemented or not according to the Metropolis rules. The temperature is initially set to a high value so that energy increases are usually allowed and local minima are escaped (Collins *et al.*, 1988). As the temperature lowers, the maximum allowable energy increase decreases. When the temperature approaches zero, the current solution should be close to the global optimum. As zero temperature is an asymptotic condition, reachable only at the limit of infinite temperature reductions, the true global optimum tends to be approached rather than reached in simulated annealing. Solutions very near the global optimum (and in discrete problems, possibly at the optimum) can be found if the algorithm runs for reasonable lengths of time (Johnson *et al.*, 1989).

Simulated annealing has been successfully applied in a wide variety of fields but it has not seen much service in the social sciences. Its main drawback is the very long run times needed by the algorithm; slow cooling of the system requires much computer time. With computers becoming increasingly fast, though, this drawback is becoming less important, and the wide applicability, relatively simple mechanics, and promising results of simulated annealing make the technique increasingly attractive.

To use simulated annealing for redistricting, Browdy suggests defining system energy as some combination of scores for a plan's compactness, contiguity, or population equality (Browdy, 1990). Other criteria may also be incorporated (see above). The procedure starts with an arbitrary initial plan.

Counties or wards are moved between districts. Each change that lowers the total energy is retained. These moves, always of only one ward at a time, represent the small, random perturbations in the system required by the Metropolis algorithm. Occasionally, moves that increase the energy are allowed to avoid the possibility of the search process being trapped in the vicinity of a local minimum. The moves continue until some stopping criterion is reached. Because the temperature can never be reduced to zero in finite time, the true global optimum may not be reached. Nonetheless, the algorithm will eventually reach a near global minimum from which no moves are accepted. After trying enough unsuccessful moves, the algorithm will halt and output the current solution. This solution can be accepted as the draft districting plan. Alternatively, the algorithm can be run a set number of times and the output with the lowest energy accepted.

The ANNEAL redistricting algorithm

To apply simulated annealing to redistricting, four things have to be decided: how solutions are represented; how solutions are perturbed; how solutions are evaluated; and what the annealing schedule is.

In ANNEAL, a possible solution or district plan is represented by a vector

$$\mathbf{x} \equiv (x_1, \ldots, x_C), \text{ where } x_c \in \{1, 2, \ldots, D\},$$

C being the number of counties and D the number of districts. This is more efficient computationally than using the vector

$$\mathbf{x} \equiv (x_1^1, \ldots, x_C^1; \ldots; x_1^D, \ldots, x_C^D),$$

where $x_i^d = 0$ or 1 $\forall i, d$.

Thus, if $x_3 = 5$, county 3 is in district 5. The numbering of the districts and counties is arbitrary.

Perturbations are carried out by moving a county to a new district. This is done by first finding all those counties that lie on a border between districts. One of these border counties is selected randomly and all the districts it touches are listed, excluding the district it belongs to. One of these touching districts is then chosen randomly and the county is flagged for transfer to that district. In numerical terms, the transfer would involve altering the integer value of one of the x_c terms. Whether transfer occurs or not depends upon the evaluation of the new solution.

After a solution is perturbed, the algorithm decides whether to accept or reject the change. The acceptance criterion is embodied in the Metropolis algorithm. If the change decreases the energy of the solution, it is always accepted (the energy level is the value of the objective function). If the energy increases, the change is accepted with probability $\exp(-\Delta E/T)$. Using the fuzzy approach in ANNEAL, the objective function may contain any number of redistricting criteria with any set of weights attached to them.

Before determining a solution's energy level, ANNEAL ensures that the membership and contiguity requirements for a solution are satisfied. That is, it checks that every district contains at least one county and that every district is composed of simply connected counties. Each county transfer only affects two districts. The district gaining the county cannot possibly become empty or noncontiguous (although it could encircle another district as noted in Section 2.4). The district losing the county could suffer both these fates. To prevent this, ANNEAL first counts the number of counties in the losing district and if the count is zero, the transfer is rejected. The algorithm then tests the losing county for contiguity. If the counties are not simply connected, the district is non-contiguous and there is no transfer. Both methods of contiguity checking – the powering of the district connectivity matrix and the counting of switching points – have been used. The latter, as one might expect, is much more efficient computationally. The remaining constraint from *Problems 1* and *2* is that every county is assigned to one and only one district. This constraint is met automatically by the new definition of the vector **x**.

If a solution passes both the membership and contiguity tests, it is evaluated to see how it performs in terms of the objective function or system 'energy'. The algorithm then determines whether or not the new solution should be accepted. The key factor is the temperature level. In simulated annealing, whilst the temperature is always lowered slowly, the starting temperature, the rate of lowering, and the number of county moves per temperature are all arbitrary. These quantities comprise the annealing schedule. Their values determine the temperature at any given moment and, therefore, determine the probability of acceptance of a given county move.

Setting an annealing schedule has been described as being as much art as science (Davis and Streenstrup, 1987). A few generalizations are possible, though. First, the initial temperature must be set much larger than the largest possible energy change (Johnson *et al.*, 1989). Doing so increases the chances of large energy increases being accepted during the early part of the algorithm's operation. Such large increases are needed to permit the algorithm to escape local minima. As the temperature decreases, however, smaller and smaller increases are allowed. Second, the rate of temperature decrease must be slow enough to permit adequate 'cooling'. If $T(s + 1) = hT(s)$, where s is the step number then it is sensible to set h to a value between 0.9 and 1.0. Values of h below 0.9 tend to lead to 'quenching' and subsequent entrapment of the algorithm in the vicinity of a local minimum. The closer h is to one, the slower the cooling and the better the results (Johnson *et al.*, 1989). Finally, the number of county moves per temperature step must be fairly large. Only then can the thermal equilibrium required by simulated annealing be achieved; once the system is at equilibrium, the temperature can be lowered again (van Laarhoven, 1988). A good practice is to accept swaps at a given temperature until $10F$ successful swaps are obtained, or $100F$ unsuccessful ones, where F

is the number of elements taking on various states, or the number of degrees of freedom present (Press *et al.*, 1986); for the districting problem, F is the number of counties, where each county can be in one of D states.

An annealing schedule also needs a stopping criterion to determine when the temperature stops being lowered. Various criteria are given in Collins *et al.* (1988). The easiest criterion to use appears to be to stop when the temperature is zero but this is approachable only as an asymptotic limit. Instead, the algorithm can stop after a fixed number of temperature decreases, after a given temperature level has been reached, or once the energy level falls below a certain value. The algorithm may also halt when no successful moves occur at a given temperature. The algorithm has then settled on a global or near global optimum and can make no further improvements to the solution. This last criterion is the most satisfactory as it stops only at the best solution for a given program run. Halting at certain cut-off values of temperature, energy, or steps leaves open the possibility of a better solution being found if only the algorithm had run longer.

In summary then, the algorithm takes a solution, and perturbs it by moving a county between districts. The new plan is tested for empty and noncontiguous districts. If any are found, the move is rejected. If none are found, the objective or energy function is evaluated, and the Metropolis algorithm decides if the swap is accepted or not. The algorithm continues testing moves at the speed set by the annealing schedule until no more are accepted. The output plan should then be at or near the global optimum.

The TransCAD–ANNEAL system

This brings us to the problem of linking ANNEAL to TransCAD. An ideal, active GIS redistricting package can best be described by imagining how a user operates the system. After entering the GIS, the user displays the state to be districted. He or she calls the redistricting algorithm, either through the keyboard or with a mouse. Any relevant inputs, such as the number of districts desired, the criteria to be used, and the weights to be attached to them, are then entered by the user. The algorithm runs and when finished displays a map of the final district plan and all output demographic data. The user may then print this map and data if desired.

As this vignette implies, all the mechanics of the algorithm are hidden from the user. Behind the screen, after the districting algorithm is called, the GIS must send the relevant input data to the algorithm. Ideally, this data is already stored in the GIS; if not, the user must be able to transfer the data directly to the GIS from a disk. The algorithm then performs the redistricting and yields an output plan with related data. The GIS now must import the algorithm's output data and display them on the screen. This two-way flow of data must be entirely independent of the user, who should be able to regard the package as a 'black box' which creates a district plan for a given state and set of criteria without needing any understanding of the algorithmic interior of the box. The user should, of course, be aware that the resulting plan is unlikely to be strictly optimal with respect to the chosen criteria and that the choice of criteria is a politico-legal and not a technical matter.

The feasibility of this ideal depends upon the particular GIS employed. Furthermore, the internal workings of the black box are also determined in part by the GIS. Subsections 5.1 to 5.10 present a description of a black box linking ANNEAL to TransCAD. Particular emphasis is put on how the GIS and the algorithm 'talk' to each other by passing data as this operation is essential to the success of the package.

The TransCAD GIS

TransCAD is a vector-based GIS developed by the Caliper Corporation of Boston, US, and is fully copyrighted by Caliper. The company's reference manual describes TransCAD as 'Transportation Workstation Software' for 'planning, management, operation, and analysis of transportation systems and facilities'. The system is designed for 'digital mapping, spatial analysis, and data retrieval for manipulation' (Caliper Corporation, 1990). The reference manual lists several potential applications of TransCAD, including transit planning, facility management, environmental impact assessment, routing and scheduling, demand modelling and forecasting, and emergency management. The description below of TransCAD is garnered from the manual and from extensive use of the system.

Options in TransCAD

When a user first enters TransCAD, he or she chooses the application to work with – a county map of the United States, for example, or a street map of a city. The geographic area is displayed on the screen along with several pull-down menus that list the options available to the user. The user can decide which database layers to display and selects one as the active layer; the layers are either point databases (cities, hospitals), line databases (roads, rivers), or area databases (counties, zip code areas). The user also can zoom in to a particular part of the map. Another option allows the user to query a particular unit in the active database. If the active layer is the county layer, the user points to a county with the mouse. The county is then highlighted on the screen, and all census data for the county is displayed.

Other screens besides the map display window are available as well. The data editor window presents all units in the active layer and all related data in tabular form and permits the user to edit and expand the database. The table editor window allows the user to enter matrix data such as intercounty traffic flows. In the charts window, pie or bar graphs can be constructed as desired, and the statistics window allows statistical analysis to be performed on given data fields. Finally, the conditions window enables the user to define logical conditions using data fields of the active layer. As an example, the user may create a condition that the number of hospitals in a county exceeds three. Conditions may be called when the user chooses another option in the map display window. If counties are active, and the hospitals condition described above is used with the select option, all counties with more than three hospitals are shaded on the screen. Furthermore, only those counties appear in the data editor window. The union or intersection of more than one condition may be used as well to find, for example, all counties with more than three hospitals and more than seventy physicians. Active layer units also can be selected by pointing at counties with the mouse, by drawing a circle around the counties, or in other ways.

Another option shades all the displayed units belonging to a given data field in conjunction with a user-defined theme. For example, if counties is the active layer again, and average household income is the data field, the user can define a theme assigning a different colour to each selected range of incomes. Each county is displayed in the colour corresponding to the income category the county falls in. Themes may also be defined in terms of line shadings instead of colours. In this way, two different themes – one colour, one shading – can be displayed simultaneously, and the relation between the two data fields can be explored.

A further powerful option in TransCAD lets the user call one of many inbuilt procedures for performing sophisticated operations. In general, the user must first run one particular procedure to create a network of nodes joined by links, based on, for example, towns linked by highways. Other procedures then may be chosen to analyse the network. Some of the procedures included in TransCAD search for the path of least cost between nodes in a network; estimate traffic flows on a network and assign flows to links; solve the travelling salesman problem for a network; evaluate spatial-interaction models; or find the optimal location of a facility on a nodal network. Because TransCAD is primarily aimed at transportation applications, the majority of procedures deal with transport networks. A few procedures are not transport-oriented and do not require a network; these procedures include one for applying nonlinear and linear models to data and one for geocoding data by zip code or other means.

An even more powerful option, the characteristic of TransCAD that led to its choice for linkage with ANNEAL, is that it permits the user to include his or her own procedures written in a programming language such as FOR-TRAN or C. By including ANNEAL as a procedure, the algorithm can be connected to TransCAD and a complete districting package created. Before describing the details of the connection, though, the method by which TransCAD executes any procedure, user-built or in-built, must be understood.

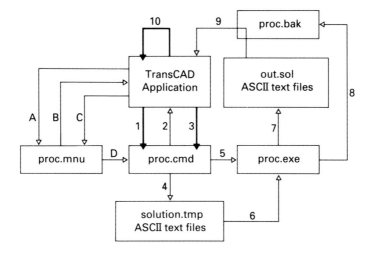

Key:

1. User calls procedure – control passes to proc.cmd
   A. user picks procedure-choose
   B. proc.mnu shows all names on screen
   C. user picks one procedure
   D. procs.mnu sends control to proc.cmd
2. Prompts in proc.cmd displayed
3. User enters info as needed
4. Proc.cmd dumps data into files
5. Proc.cmd calls proc.exe
6. Proc.exe imports input data and executes
7. Proc.exe exports output data
8. Proc.exe calls proc.bck
9. Proc.bck controls import of output data into TransCAD database
10. User modifies and displays as desired.

Figure 11.7. Diagram of the relationships between the elements of TransCAD-ANNEAL. The thick lines indicate user's actions. The activities represented by thinner lines are hidden from the user.

Execution of procedures in TransCAD

When a user wishes to execute a procedure in TransCAD, he or she uses the mouse to select the procedure-choose option. A list of procedure classes are then given. After the user picks one class, all procedures in that class are displayed. The user selects the desired procedure; the names of all the procedures are stored in the file procs.mnu. Once the procedure is selected, the procs.mnu file starts the procedure execution along the path shown in Figure 11.7.

Selecting a procedure causes the procs.mnu file to pass control to a command directive file containing instructions for sending the required data from TransCAD to the actual program embodying the procedure. The directive file, of the form proc.cmd, prompts the user on the screen for any required input parameters. The user enters these values at the keyboard, and the directive file stores them in a file named solution.tmp. The proc.cmd file next transfers any needed input data from the TransCAD database to various ASCII format, comma-delimited text files. The algorithm code is then called.

The program code for the procedure, stored in the file proc.exe, executes as if in a DOS operating system. During execution, the procedure imports input data from the ASCII text file and solution.tmp file and outputs results and data to a solution file such as out.sol and to other ASCII format, comma-delimited text files. When the code halts, control is passed to the proc.bck file.

This file contains the back-end command directives for sending output data back into TransCAD. The file takes the out.sol and output ASCII text files and sends their contents to the appropriate TransCAD database. The user may then view the data in the data editor window or use the data as a basis for selecting by condition or displaying by theme. Thus, the entire procedure execution is complete.

Connecting ANNEAL to TransCAD

With the above explanation of TransCAD procedure execution finished, the details involved in connecting ANNEAL can be presented. There are four steps: creating the anneal.exe file, creating a contiguity file, modifying the procs.mnu file, and creating the anneal.cmd and anneal.bck files. The first two files must be created outside TransCAD entirely, whilst the last two files can be created within the TransCAD directory.

Before continuing, a mention of the personal computer which ran TransCAD is needed. The work described here was performed on a Viglen SL3-SX with 8 Mb of RAM, a math co-processor and an 80Mb hard disk, on which was mounted the DOS operating system and editor. As ANNEAL is written in FORTRAN, a FORTRAN compiler was also installed. The hard disk was partitioned into two drives, C and D. TransCAD was installed on the C drive.

Creating the anneal.exe file

Because TransCAD has no programming language compiler in it, it can only run procedures stored as .exe files, which are already compiled. Thus, the ANNEAL algorithm has to be entered first into the PC as a .for file then compiled to produce an .exe file. This file is stored in the TransCAD file directory, in the Procs subdirectory, which contains all the .exe, .cmd, and .bck files.

Creating the contiguity file

For the first of the two methods of contiguity checking described earlier, ANNEAL requires a connectivity matrix. This matrix could conceivably be created in the table editor window of TransCAD and then exported to ANNEAL. However, entering every value of such a table by hand is unnecessarily tedious so another FORTRAN program is used to create the contiguity file. This program merely requires the user to enter the counties touching each county. All the zero values for non-touching counties do not have to be entered. Furthermore, each pair of touching counties has to be entered only once, not both ways, as the table editor requires. The program outputs the values to the file gmcont.dat in ASCII format. The gmcont.dat file is then stored in a file directory called Gisdata, located on the D drive. Ideally, the connectivity data should be obtained directly from TransCAD locational data files but we have not been able to do this.

The ordering of the counties in the file, a minor consideration when running ANNEAL on a VAX, takes on greater significance in TransCAD. The counties in TransCAD databases are stored in a particular order reflecting the system's use of quadtrees to organize area data. Thus, the demographic data is kept in the same order. This order must be used in the contiguity file to preserve consistency. To find the TransCAD county ordering, one must select the counties in the state of interest. The data editor then displays the counties in the TransCAD order.

Altering the procs.mnu file

The procs.mnu file, stored in the TransCAD file directory, contains the names and classes of procedures in the procedures menu. To include a new procedure, the following format is used:

>[class]
><class name>
>procs\\<algorithm name>.cmd, '<procedure name>'.

Thus, for the ANNEAL algorithm, the format is:

>[class]
>Districting Procedures
>procs\\anneal.cmd, 'Districting by Simulated Annealing'.

This causes the districting procedure to appear in the procedures menu in the map display window.

Creating the command directive files

The command directive files, stored in the Procs subdirectory, contain commands for transferring data between TransCAD and ANNEAL. The TransCAD reference manual contains a full list of the commands available; only the commands used for ANNEAL are explained here.

Figure 11.8 shows the anneal.cmd file. The two 'set' commands save the databases and screen display before execution to prevent their corruption; both are restored after the procedure ends. The two 'show' commands display two messages on the screen. These messages warn the user to ensure all counties in the state are selected and the contiguity file GMCONT file is available.

The next four 'get' commands ask for information from the user: the output solution file name, the number of counties, the number of districts desired, and the contiguity file name. These inputs are stored in solution.tmp for access by the anneal.exe file. The two 'dump' commands place the database (db) name in solution.tmp and export the FIPS, Population, and District data fields for the county base to an ASCII format, comma-delimited text file pop.dat. These fields contain, respectively, the FIPS or Census Bureau identification code for each county, the population of each county, and the number of the district each county is assigned to in the initial solution. Finally, the 'run' command calls the anneal.exe file.

The anneal.bck file is in Figure 11.9. Only one command is needed; the 'import' directive imports the ANNEAL output data into TransCAD. These output data are stored in the ASCII format, comma-delimited text file district.txt and consist of each county's FIPS and District data field values. Again, the FIPS value is the county identification code, and the District value is the number of the district to which the county is assigned in the final district plan. These values are directly transferred into the corresponding fields in the county database. Because district.txt holds all output data needed, nothing is stored in the solution file. Thus, anneal.exe creates an empty solution file, out.sol, so TransCAD can recognise the existence of the file and then carry on with procedure execution.

Using the TransCAD-ANNEAL package

Once the above four tasks are complete, a user can treat the districting procedure as a black box. After making sure all the required demographic and other data are in TransCAD databases, the user selects all the counties in the state of interest and enters the data editor window. There, the input district plan is entered in the District data field by entering the number of the district to which each county is initially assigned. After returning to the map display window, the user chooses 'Districting by Simulated Annealing' from the procedures menu. The warning messages appear in turn; the user must hit a key to clear them. The series of prompts are then displayed, causing the

```
set execution condition close database on;
set execution condition screen save on;

show message 'Make sure you have selected all desired counties.'
show message 'Make sure the appropriate contiguity file is available.'

get filename solution prompt 'Enter a solution filename';
get integer prompt 'Enter the number of counties';
get integer prompt 'Enter the number of districts';
get string length 30 prompt 'Enter the name of the contiguity file (in single
    quotes)';

dump db name;
dump selected attributes to pop.dat 3 'FIPS' 'Population' 'District';

run procedure 'procs\\anneal' using procs\\anneal.bck;
```

Figure 11.8. The anneal.cmd file.

```
import file c:\transcad\district.txt;
```

Figure 11.9. The anneal.bck file.

user to enter the requested information. The procedure executes, and when finished, the screen informs the user that the output data is being imported into the TransCAD database. When this data transfer is done, the user may display the counties by theme so each district is shown in a different colour. Once the user is finished analysing the output, he or she can rerun the procedure or exit TransCAD. After exiting the GIS, the user should delete all temporary files created by the execution of the FORTRAN code.

Results obtained from the TransCAD-ANNEAL package

Running the above package quickly revealed that the run times of ANNEAL with the powered connectivity matrix used in the contiguity check were intolerably long on the PC. Even on a VAX they were far from speedy. The PC redistricting package was first run on Louisiana with population equality as the single optimisation criterion. After five days, the program was still executing and had to be stopped!

To demonstrate the algorithm did work within TransCAD, it was applied to New Hampshire, a state with only ten counties and two districts. The procedure ran for fifteen minutes and produced an intuitively appealing map when execution ended. The algorithm was also run for Maine, a state with sixteen counties and two districts. After a run time of two hours, another acceptable map was produced. In both cases, the population deviations were quite tolerable. The long run times, though, suggested that other approaches were needed.

Improvements

One possibility is to try to streamline ANNEAL. This could be done by making the search procedure more guided than random, whilst preserving the possibility of making moves which take the search back 'uphill' to avoid local minima. This has not been tried as yet. Another possibility, in principle, is to link the micro running TransCAD to a mainframe to get the latter to perform the heavy-duty numerical work. In practice, it appears that this cannot be done in any integrated way.

The strategy we chose to explore was to dispense with the time-consuming contiguity check and use some surrogate means for achieving a contiguous plan. In particular, we tried various measures of proximity between individual counties and attempted to maximise the aggregate proximity score for the plan. The general form of the problem we attempted to solve was as follows:

Problem 3

maximize $\sum_d \sum_{ij} g_{ij} x_i^d x_j^d$

s.t.

$$\sum_i x_i^d \geq 1 \qquad\qquad \forall d$$

$$\sum_d x_i^d = 1 \qquad\qquad \forall i$$

$$(1 - \varepsilon)\, \text{P/D} \leq \sum_i p_i x_i^d \leq (1 + \varepsilon)\, \text{P/D} \qquad \forall d$$

$$x_i^d = 0 \text{ or } 1 \qquad\qquad \forall i,d$$

where g_{ij} is the proximity of counties i and j and ε is a number between zero and 1 (so the third constraint restricts the population of each district to some specified range of values around the target population level P/D). Various proximity measures were tested. These included the value of the corresponding k_{ij} term in the connectivity matrix (0 or 1), and the value of the corresponding term in the nth power of the connectivity matrix, where n was a small positive integer. Other measures of geographical or economic proximity could have been tried as well.

It was for the solution of this problem that we turned to the NAG routines referred to at the start of the paper. *Problem 3* would be a quadratic programming problem if there was no integer restriction on the variables. NAG has a prototype mainframe qp algorithm with integer constraints that we were given access to for some trial runs but there is no micro version of this program. Thus, we attempted to solve *Problem 3* without the integer constraint which yielded, at best, some near-integer, near-contiguous plans. To progress further with this line of attack, it would be necessary to develop a modified version of a qp routine of our own or to amend the NAG code to include an integer restriction. If we could do that though, we could equally well add a contiguity check. We would still face the problem that the presence of that check implicitly introduces the possibility of non-global local optima. We might, however, decide to live with that possibility and devise a system requiring multiple runs from different starting points in an attempt to circumvent it.

Clearly, much more needs to be done both on the algorithms and on their relationship to the GIS but it is equally clear that we have an active redistricting system that works, albeit painfully slowly on the rather low-powered machine we used for the development work. It would be a significant step forward if the GIS allowed us to obtain connectivity data directly from its databases and, subsequently, allowed us to query the districts created in the plan in the same way that their constituent counties can be queried. If it is possible for either of these things to be done in TransCAD, we have not discovered how to do it. Finally, there would be some benefit (though considerable computational time cost) in being able to produce dynamic displays of possible district plans as they are generated by the algorithm, or at least to display a sample of such plans. This facility would be of even greater value in connection with some of the dynamic simulation work being pursued on other related projects.

References

Browdy, M., 1990, Simulated annealing – an improved computer model for political redistricting, *Yale Law and Policy Review*, **8**, 163–179.

Cerny, V., 1985, Thermodynamical approach to the travelling salesman problem: an efficient simulation algorithm, *Journal of Optimization Theory and Its Applications*, **45**, 41–45.

Collins, N., Eglese, R. and Golden, B., 1988, Simulated annealing – an annotated bibliography, *American Journal of Mathematical and Management Sciences*, **8**, 209–308.

Davis, L. and Streenstrup, 1987, Genetic algorithms and simulated annealing: an overview, L. Davis (Ed.), *Genetic Algorithms and Simulated Annealing*, London: Pitman.

Economist, The, 1992, Congressional redistricting: no grey areas, 8 February 1992, pp. 48–49.

Gearhart, B. and Liittschwager, J., 1969, Legislative districting by computer, *Behavioural Science*, **14**, 404–417.

Johnson, D., Aragon, C., McGeoch, L. and Schevon, C., 1989, Optimization by simulated annealing: an experimental evaluation; part 1, graph partitioning, *Operations Research*, **37**, 865–892.

Kirkpatrick, S., Gelatt, C. and Vecchi, M., 1983, Optimization by simulated annealing, *Science*, **220**, 671–680.

Metropolis, N., Rosenbluth, A., Rosenbluth, M., Teller, A. and Teller, E., 1953, Equation of state calculation by fast computing machines, *Journal of Chemical Physics*, **21**, 1087–1091.

Nagel, S., 1965, Simplified bipartisan computer redistricting, *Stanford Law Review*, **17**, 863–899.

Pierce, T. and Macmillan, W., 1994, *Fuzzy Optimization and the Redistricting Problem*, forthcoming.

Press, W., Flannery, B., Teukolsky, S. and Vetterling, W., 1986, *Numerical Recipes*, Cambridge: Cambridge University Press.

Rossiter, D. and Johnston, R., 1981, Program GROUP: the identification of all possible solutions to a constituency delimitation process, *Environment and Planning A*, **13**, 321–238.

van Laarhoven, P. J. M., 1988, *Theoretical and Computational Aspects of Simulated Annealing*, Centre for Mathematics and Computer Science, Amsterdam.

Vickery, W., 1969, On the prevention of gerrymandering, *Political Science Quarterly*, **76**, 105–110.

Wilson, A. G., Coelho, J. D., Macgill, S. M. and Williams, H. C. W. L., 1981, *Optimization in Locational and Transport Analysis*, Chichester: John Wiley.

12

A surface model approach to the representation of population-related social indicators

Ian Bracken

Introduction

Census data have long been used in the generation of various indices of deprivation based on the concept of social indicators (Hatch and Sherrott, 1973; Carlisle, 1972; Smith, 1979). Many such indices have become firmly integrated in government policy making processes at both national and local levels and particularly in fields such as housing, regional policy and health provision (Jarman, 1983, 1984; Senior and McCarty, 1988). Many widely used indicators are of a 'simple' nature, for example, the proportion of dwellings regarded as 'overcrowded' (in British terms this means households having more than 1.0 or 1.5 persons per habitable room); the proportion of dwellings lacking basic amenities (inside WC or fixed bath and hot water system); the percentage of persons unemployed; the percentage of households not owning a vehicle; the proportion of single parent families, and so on. Most of these are conventionally derived from the decennial population censuses and are normally computed at the ward level as the spatial unit. This is a local administrative unit in the UK of between 5,000 and 20,000 persons. In recent years, more complex indicators have become common in policy advice contexts using not only multivariate census data (using 'weights' to represent the relative importance of one census variable against another), but also incorporating data from other sources. Of these, the most important is anonymised address-based information derived from routine administration, such as health care (i.e. visits to doctors, prescriptions, claims for benefits, and so on). The increasing use of information technology in such administration has greatly expanded the volume of such data (Hays et al., 1990) and its potential for spatial analysis particularly as much data is now routinely postcoded.

Although there is considerable debate both about the philosophical and methodological nature of indicators (Talbot, 1991), in practice they are extensively used to make judgments about a wide range of social and economic conditions, both directly and indirectly. An example of the latter is the level and distribution of car ownership (a census variable) used as a proxy for household income. The advantages of indicators based on a census are that the data are available as a regular series; the definitions of variables used are well understood, though not always consistent from census to census; the data are readily available in machine readable form; the coverage of Britain (for example) is spatially complete; and the data are spatially comprehensive in that all zones have the full set of data 'counts'. There are however some long standing and often overlooked difficulties. These concern mainly the spatial arrangement of the zones by which censuses are conducted. In Britain, these zones (called Enumeration Districts – EDs) ensure complete coverage of all the land and number just over 130,400. They vary greatly in size and shape however, and even in urban areas are barely homogeneous. As their practical purpose is to facilitate enumeration, it is often the case that their boundaries fail to conform to any social distinction in spatial terms, such as types of housing. Wards typically comprise 20 or 30 enumeration districts in urban areas, though often many fewer in rural areas.

The EDs are also the basic unit for which data are published involving the assembly of information for between 50 and 450 persons. In rural areas, these zones can be large, and therefore may include several distinct settlements. More generally, in both urban and rural areas the zones will include areas of 'un-populated' land, such as commercial areas and farmland. In addition, the spatial definitions of the zones often change from census to census, and they by no means conform to other functional areas used for example for employment recording, health service delivery, community planning purposes and so on. In general, whilst the zones form an efficient framework for enumeration (which in the UK is done by individual household visit) and become the basic spatial unit for data provision in order to preserve confidentiality, they create many practical and widely acknowledged problems for spatial analysis (Openshaw, 1984; Unwin, 1984; Flowerdew and Green, 1989).

It is important to note that in using census-type data, there are both practical issues of data display and underlying theoretical issues of data representation present (Rhind, 1983; Peuquet, 1988). For practical purposes, standard choropleth techniques are frequently used to display data at the basic level for a local area. This might show for example the proportion of sub-standard dwellings using a census-derived criterion and a typical display may well take the form of shading on a zonal map base. Such mapping is now very readily produced by many PC-based GIS and this availability makes all the more urgent a careful assessment of the 'quality of the message' that is provided. Such displays can be highly misleading, mainly because the 'fixed' area base has no relationship to the distribution of the phenomenon in question; the apparent distributional properties of the topic in question will be distorted by the imposed boundaries; and all areas must be populated, which is clearly a misrepresentation at any scale. If spatial policy and the allocation of resources takes its advice directly from such displays at the local level, then misinterpretation is very likely. As the data values associated with each ED are area aggregates, the actual values may be as much a product of the zonal boundaries and their locations as a representation of an underlying geographical distributions. This has clear implications for advice in the effective allocation of public resources. The underlying issue is however more theoretical and concerns the most appropriate way in which population and hence population-related phenomena, *should* be represented in a spatial data structure, and hence be portrayed by graphic and cartographic means (Forbes, 1984; Langford *et al.*, 1990; Trotter, 1991).

On the subject of the integration of data from different sources, it can be argued *a priori* that broadening the information base for the generation of spatial indicators by such integration should lead to more meaningful spatial indices. An integration of census-type and address-type data offers many possibilities, though there are some difficulties due to the fact that the 'area' properties of census data and the 'point' properties of address-coded data are both approximations of underlying spatial distributions. As it is unlikely that analysts outside particular organizations can have access to 'non-anonymised' personal household records, it is increasingly common for the confidentiality of household data to be maintained by making information available only at the postcode level. In the UK, a postcode unit (i.e. the most local level) relates to about 15 dwellings, or about 40 persons, and a geographical reference to 100 metres accuracy is available in machine readable form for all 1.5 million codes.

Technically, the integration of census (area) and postcode (point) data for these purposes can be done in three ways: first, point address locations may be allocated to census zones so that in effect address data becomes another aggregate field of the zonal record; second, each address location can be assigned data from its enveloping zone, so that the point takes on the attribute of its surrounding area; or third, both types of data can be re-represented geographically onto a 'neutral' base in the form of a georeferenced grid. The limitation of the first two possibilities (which incidentally are often provided by GIS functionality) is that the inherent assumptions about the spatial distribution of the population 'within the zones' and 'around the points' are not being fully acknowledged and the attendant errors are rarely evaluated. The third alternative, which is advocated here in the form of a surface model, would appear to offer some important advantages. This can exploit the recognized properties of a georeferenced grid as a data structure for spatial analysis (Kennedy and Tobler, 1970; Tobler, 1979; Browne and Millington, 1983).

A surface model

It can be argued that the representation of population and population-related phenomena, which are punctiform in nature, can best be done using a structure which is independent of the peculiar and unique characteristics of the actual spatial enumeration. One way in which this can be achieved is to transform zone-based census data and point-based address data into a data structure that represents population distribution more plausibly as something approaching a continuous surface. To operationalize this approach, it requires a change in the basic assumption of data representation, namely that the data value (i.e. the count) is seen not as a property of an area or a point, but rather a property of a location at which the population (and related phenomena) are distributed. It is then further assumed that useful information about this distribution can be derived from a local, spatial analysis of a *set* of the basic area or point data values. Using this assumption, a model can be created to generate a spatial distribution of population (or households) as a fine, and variable resolution geographical grid and with assessable error properties (Bracken and Martin, 1989).

As already noted, in the UK, machine readable georeferences are available for both postcode units and census zones, all related to the Ordnance Survey National Grid. For the former, a single file contains all residential unit postcodes and their grid locations to a 100 metres approximation. In practice, this means that a small number of codes share a common grid reference. In the case of the census, every ED has assigned to it manually a georeferenced centroid which is population-weighted, that is located at the approximate centre of the populated part of the zone. Again, this information is a 100 metres approximation. The spatial pattern of census centroids or postcode locations will be quite different for an isolated settlement, a village, a linear settlement, a suburb or the central part of city. One such location cannot say anything about a local distribution of population, but a kernel-based analysis can be used to generate a plausible distribution according to the local density of a given set of data locations. This analysis uses a derivative of the moving kernel density estimation technique (Bowman, 1973; Silverman, 1986; Martin, 1989) and can be shown to be capable of both reconstructing settlement geography from the georeferenced population 'points', and of preserving the volume (i.e. the count) under a density surface.

In the analysis, a window is positioned over each data point in turn, and it is the size of this window which is made to vary according to the local density of points around that location. In effect this provides an estimate of the size of the areal unit assumed to be represented by the current location, whether census centroid or postcode. The count associated with this location is then distributed into the cells falling within the window, according to weightings derived from a distance-decay function, as used by Cressman (1959). This function, which here uses a finite window size, is formulated to increase the weight given to cells nearest the data point in relation to local point density (Martin and Bracken, 1991). The allocation involves only integer (i.e. whole person) values which means that the distribution process has finite, and in practice quite local bounds thereby reconstructing a good estimate of populated and unpopulated areas.

In summary, the assignment to each cell in the grid is obtained as:

$$P_i = \sum_{j=1}^{c} P_j \cdot W_{ij}$$

where P_i is the estimated population of cell i, P_j is the empirical population recorded at point j, c is the total number of data points, and W_{ij} is the unique weighting of cell i with respect to point j. Cell i will not receive population from every point location but only from any points in whose kernel it falls (W_{ij} will be zero for all other cell-point combinations). At each location, the kernel is initially defined as a circle of radius r, and is locally adjusted according to the density of other points falling within distance r of that location j. Weights are then assigned to every cell whose centre falls within the adjusted kernel, according to the distance decay function:

$$W_{ij} = f\left(\frac{d_{ij}}{r_j}\right)$$

where f is the distance decay function; d_{ij} is the distance from cell i to point j, and r_j is the radius of the adjusted kernel defined as:

$$r_j = \frac{\sum_{l=1}^{k} d_{jl}(j \neq 1)}{k}$$

where $l = 1, 2, \ldots, k$ are the other points in the initial kernel window, of radius w. If there are no other points with $d_{ij} < w$ (i.e. $k = 0$), then the window size w_j remains set equal to w. This sets the dispersion of population from point j to a maximum in the case where there are no other points within the entire window. In effect, the value r determines the maximum extent to which population will be distributed from a given location. Where $k > 0$, a greater clustering of points around point j (smaller radius r_j) will result in a smaller window and hence greater weights being assigned to cells falling close to j. In the most localized case, where r_j is less than half the size of a cell, it is probable that the cell containing the location will receive a weight of 1, and the entire population at that location will be assigned to the cell. It can also be noted that the weight given to a cell centre spatially coincident with the location j also cannot be greater than 1, as the location must then contribute population to the output grid exactly equivalent to its attribute value. Once weights have been assigned to every cell in the kernel, they are re-scaled to sum to unity thereby preserving the total population count P_j to be distributed from location j. As some cells fall within the kernels based on several different points, they receive a share of the population from each. In contrast, other cells, beyond the region of influence of any points, remain unvisited and receive no population and hence define the unpopulated areas in the complete raster surface database.

In the early implementations of this model (Martin, 1989) a simple distance decay function was used in which the weighting was directly and inversely related distance from each input location. Latterly, a more flexible alternative has been developed to define the weightings as:

$$w_{ij} = \left(\frac{w_j^2 - w_{ij}^2}{w_j^2 + d_{ij}^2} \right)^{\alpha} (d_{ij} < w_j)$$

This function has a value of zero when d_{ij} is greater than or equal to r_j. Increasing the value of alpha increases the weight given to cells close to the data point thereby achieving greater control over the form of the distance decay.

Using these techniques, a highly plausible raster model has been implemented capable of providing a good approximation to a spatially continuous estimate of population from census or postcode sources. The importance of being able to assess the errors of the process should not be overlooked (Goodchild, 1989). Evaluating the allocation of populated and non-populated cells from the British census model at 200-metre (i.e. fine) cell resolution against a rasterization of digitized residential areas from large-scale maps, showed that in the case of the city of Bristol, less than six per cent were wrongly designated as 'populated or not' mainly on urban fringes. Moreover, re-estimation of the population from the model database by digitization and overlay of census zones, produced estimates of the 'real' values with a mean error of less than four per cent of the population at a 95 per cent confidence level. Although the model appears to work well with British census and postcoded data, nevertheless it is recognized that all processes involving kernel-based estimation will be sensitive to the bandwidth of the kernel for which evaluation techniques have more generally been suggested (Bowman, 1964).

Using this model a series of population-related raster databases have been generated from zone-based 1981 British Census data at fine resolution and for extensive areas. National 'layers' in this database using a 200-metre cell size comprise 330,000 populated cells out of the 16 million cells in the overall national grid. A number of 'key' census variables have been modelled onto the same grid, and spatial indicators can readily be computed at the level of individual cells using the population (or household) surface as the denominator. As the density of postcode locations is greater than the density of census zone centroids, the model will support a much finer resolution grid where data are based on the unit postcodes. It should be noted that the two data sources can be used in conjunction, for example the postcode locations can be used to generate a finer spatial distribution of the census zone population counts. In this case the census provides the counts and the postcodes provide the distributional information via the model.

Application to spatial indicators

Possible applications of these techniques to policy making are varied but can be briefly illustrated in this paper by reference to some typical surface products drawn from a national 'indicator surface' database. For this paper these focus on the Cardiff and Bristol regions, though others have been illustrated elsewhere (Bracken, 1989; 1991). A key indicator for social and economic analysis is that of population change, and Figure 12.1 shows the model output using 200-metre cell resolution. Here the cells been classified for the purposes of display using just three intervals, namely cells that experienced a loss of over 15 persons during or more over the inter-censal period; those whose population change only varied between plus and minus 15 persons; and cells that experienced a gain of over 15 persons. The power of the method to discriminate quite small changes in a variable at the local level is well illustrated by this, even though the illustration is taken directly from the national coverage at this resolution. The movement of the population from the 'inner' parts of the city to new, planned housing areas is clearly revealed, as similarly is an east-west band of new development following the route of a motorway. The visualization of the 'texture' of population change revealed in this way can be contrasted to the much less informative portrayal using zonal representation. Moreover the clustering of areas into classes is entirely data dependent and not a function of imposed boundaries.

Figure 12.2 shows the same approach for the representation of unemployment (using the census definition of 'seeking work'), here plotted into four classes. Again the discrimination within the city is clearly revealed, as are the unemployment blackspots in the industrial 'valley' areas to the north. This figure also reveals clearly the way in which the model has reconstructed the settlement pattern in terms of populated and un-populated cells from the overall coverage of the census zones. Figure 12.3 shows a 'housing conditions' surface using the standard British indicator of 'dwellings lacking basic facilities' while Figure 12.4 shows a measure of 'overcrowding' computed from the counts of household numbers and the number of habitable rooms that they occupy. These examples all invite comparison with the contemporary use of ward or area based mappings of such indicators. However, a fine resolution cell based approach allows new local areas to be readily defined using simple cell classification and aggregation (or clumping) techniques.

Figure 12.5 illustrates the use of the surface concept to integrate information from postcodes and the census. This shows an index of 'access to General Practitioner' in part of the County of Avon, including the city of Bristol. The locations of GPs are postcoded from which a service provision surface is generated. This is related at the level of individual cells to total population as a ratio, thereby generating directly an accessibility index. Such product surfaces can readily be generated using for example, other variables such as population age groups, or social class. Moreover, a direct evaluation can be made of the sensitivity of a distribution to different combinations of subgroups within an overall population.

In general the fine resolution of the indicator surface for policy purposes can be noted, together with its temporal and spatial consistency properties. Moreover, as the data are in the form of a surface database the method lends itself to two possibilities for further development. First, extensive applications

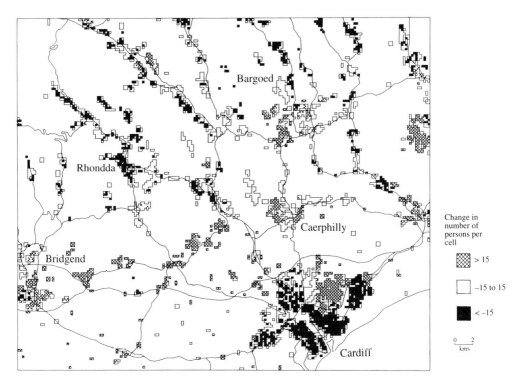

Figure 12.1. 200-metre raster surface modelled absolute population change 1971–81 for the Cardiff region.

through automation are possible given the machine readable sources of all information, whilst preserving fine resolution for interpretation. This means that larger scale, 'reference' surfaces can be created to assist in the relative interpretation of local indicator surfaces. Second, the data can be used analytically for a range of spatial analyses using standard GIS functions such as neighbour classification, accessibility indices related to distance, i.e. the grid, and so on.

Conclusions

A recurrent theme of government is that public resources should be used wisely, and be devoted to the amelioration of problems in areas with the 'greatest need'. Experience suggests that resources are not always channelled to serve those in greatest need and there are important political considerations too that have to be recognized. However, geographical analysis has an important role to play in the allocation of public resources and it is highly desireable that it should serve this need with validity. The work described here attempts to contribute to this in two ways.

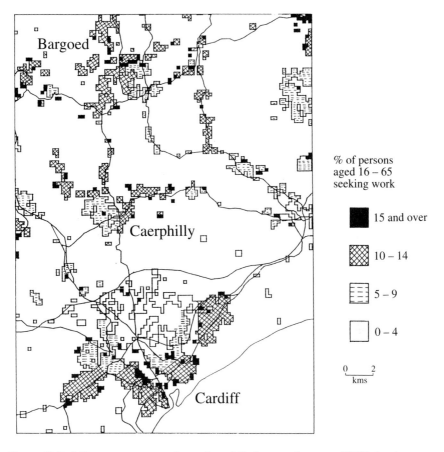

Figure 12.2. 200-metre raster surface of modelled unemployment (1981 data) distribution for the Cardiff region, plotted in four classes.

The first is to provide a form of data representation that is inherently more suitable to the display, manipulation and portrayal of socioeconomic information. This can be looked at from the point of view of 'adding value' to existing census and census-type data. Such representation leads directly to a highly disaggregate spatial visualization through 'indicator surfaces'. One advantage is that the errors present are well understood and can be explicitly evaluated in contrast to other techniques for approximating distributions within zones. The method allows the direct portrayal of areas of 'greatest need' by ready manipulation of data through classification and 'cut-offs'. The inherent spatial flexibility is vital for it allows regional and local comparison, without restriction, to provide the essential context within which the judgements about the 'meaning' of an indicator must be made for effective local policy justification and implementation.

Second, the 'relational' properties of the surface and its ability to integrate

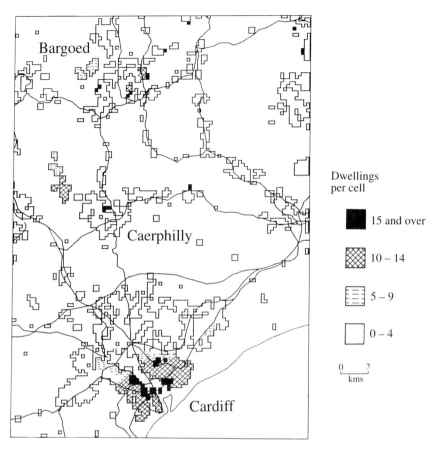

Figure 12.3. 200-metre raster cell surface of dwellings lacking basic facilities (inside toilet and fixed bath) showing cells classified by absolute numbers.

data from different sources means that the delivery of policy can be evaluated against need. For example, the recipients of housing improvement grants (as shown by their postcodes) or medical treatment can be directly related to an indicator surface. In effect, a surface of 'delivery' may be compared to a surface of 'need'. Although the respective data will be drawn from different processes (e.g. census and administrative/postcode) and have different spatial forms, the surface concept facilitates their effective integration under known conditions. Indeed there is a wide range of geographic data that can be brought together in this way, including for example address/site-coded registers of planning applications to develop land and satellite observations of land use change.

This chapter has attempted to demonstrate the versatility of the surface data model concept applied to areas of public policy. The method outlined can be seen to address a central requirement in the use of social and economic

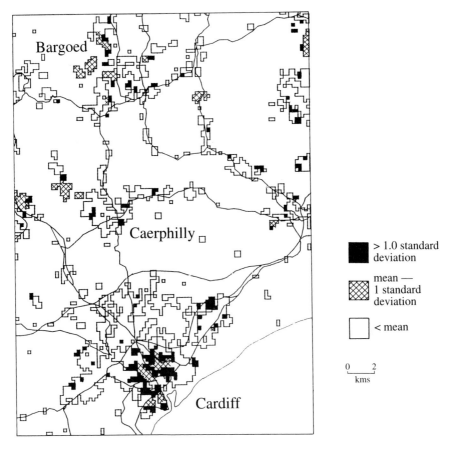

*Figure 12.4. 200-metre raster cell surface of overcrowded dwellings (OPCS defini-
tion) – >1 person per habitable room – showing cells above the mean, and >1
standard deviation above the mean for the region.*

data derived from a variety of sources in informing public policy-making,
namely to ensure that the form of representation does accord as closely as
possible with the nature of the phenomenon. More generally, the approach
has important properties of data integration and analysis that are likely to
have wider application for information handling within geographical analysis.

References

Bowman, A. W., 1973, A comparative study of some kernel-based nonparametric
    density estimators, *Journal of Statistical Computation and Simulation*, **21**, 313–327.
Bowman, A. W., 1984, An alternative method of cross-validation for the smoothing
    of density estimates, *Biometrika*, **71**, 353–360.
Bracken, I., 1989, The generation of socioeconomic surfaces for public policy making,
    Environment and Planning B, **16**, 307–326.

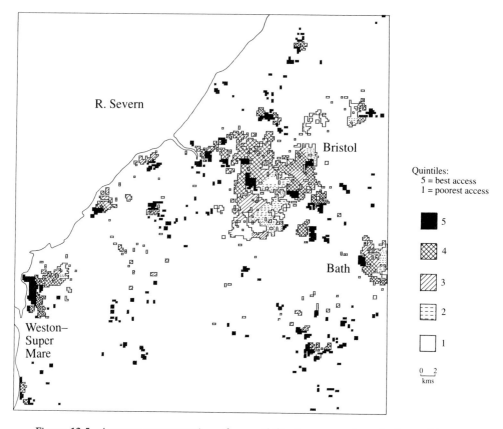

Figure 12.5. A raster representation of accessibility to a general medical practitioner plotted in five classes for the Bristol region using 200-metre cells.

Bracken, I., 1991, A surface model of population for public resource allocation, *Mapping Awareness*, **5**(6), 35–39.

Bracken, I. and Martin, D., 1989, The generation of spatial population distributions from census centroid data, *Environment and Planning A*, **21**, 537–543.

Browne, J. J. and Millington, A. C., 1983, An evaluation of the use of grid squares in computerised choropleth maps, *The Cartographic Journal*, **20**, 71–75.

Carlisle, E. 1972, The Conceptual Nature of Social Indicators, Schonfield, A. and Shaw, S. (Eds.), *Social Indicators and Social Policy*, London: Heinemann.

Cressman, G. P., 1959, An operational objective analysis system, *Monthly Weather Review*, **87**(10), 367–374.

Flowerdew, R. and Green, M., 1989, Statistical methods for inference between incompatible zonal systems, *Accuracy of Spatial Databases*, M. F. Goodchild and S. Gopa (Eds.), London: Taylor & Francis, pp. 239–248.

Forbes, J., 1984, Problems of cartographic representation of patterns of population change, *The Cartographic Journal*, **21**(2), 93–102.

Gatrell, A. C., Dunn, C. E. and Boyle, P. J., 1991, The relative utility of the Central Postcode Directory and Pinpoint Address Code in applications of geographical information systems, *Environment and Planning A*, **23**, 1447–1458.

Goodchild, M. F., 1989, Modelling error in objects and fields, *Accuracy of Spatial Databases*, M. F. Goodchild and S. Gopal (Eds.), London: Taylor & Francis, pp. 107–114.

Hatch, S. and Sherrott, R., 1973, Positive discrimination and the distribution of deprivations, *Policy and Politics*, **1**, 223–240.

Hays, S. M., Kearns, R. A. and Moran, W., 1990, Spatial patterns of attendance at general practitioner services, *Social Science Medicine*, **31**(7), 773–781.

Jarman, B., 1983, Identification of underprivileged areas, *British Medical Journal*, **286**, 1705–1709.

Jarman, B., 1984, Underprivileged areas: validation and distribution of scores, *British Medical Journal*, **289**, 1587–1592.

Kennedy, S. and Tobler, W. R., 1983, Geographic interpolation, *Geographical Analysis*, **15**(2), 151–156.

Langford, M., Unwin, D. J. and Maguire, D. J., 1990, Generating improved population density maps in an integrated GIS, *Proceedings of the first European Conference on Geographical Information Systems*, EGIS Foundation, The Netherlands Utrecht.

Martin D., 1989, Mapping population data from zone centroid locations, *Transactions of the Institute of British Geographers,* New Series, **14**(1), 90–97.

Martin, D. and Bracken, I., 1991, Techniques for modelling population–related raster databases, *Environment and Planning A*, **23**, 1069–1075.

Openshaw, S., 1984, *The Modifiable Areal Unit Problem*, Concepts and Techniques in Modern Geography No. 38, Norwich: Geo Books.

Peuquet, D. J., 1988, Representations of geographic space: toward a conceptual synthesis, *Annals of the Association of American Geographers*, **78**, 375–394.

Rhind, D., 1983, Mapping Census Data, Rhind, D. (Ed.), *A Census User's Handbook*, London: Methuen, pp. 171–198.

Senior, M. L. and McCarthy, T., 1988, Geographical approaches to health and deprivation, *Medical Geography: Its Contribution to Community Medicine*, L. P. Grime and S. D. Horsley (Eds.), Western Regional Health Authority, Manchester.

Silverman, B. W., 1986, *Density estimation for Statistics and Data Analysis*, Routledge.

Smith, D. M., 1979, The identification of problems in cities: applications of social indicators, *Social Problems and the City*, D. T. Herbert and D. M. Smith (Eds.), Oxford: Oxford University Press, pp. 13–32.

Talbot, R. J., 1991, Underprivileged areas and health care planning: implications of use of Jarman indicators or urban deprivation, *British Medical Journal*, **302**, 383–386.

Tobler, W. R., 1979, Smooth pycnophylactic interpolation for geographic regions, *Journal of the American Statistical Association*, **74**, 519–530.

Trotter, C . M., 1991, Remotely-sensed data as an information source for geographical information systems in natural resource management: a review, *International Journal of Geographical Information Systems*, **5**(2), 225–239.

Unwin, D., 1981, *Introductory spatial analysis*, London: Methuen.

13

The Council Tax for Great Britain: a GIS-based sensitivity analysis of capital valuations

Paul Longley, Gary Higgs and David Martin

Introduction

During the last five years, Britain has experienced a rapid succession of methods of local taxation. The redistributive consequences of these changes in socio-economic terms are now broadly understood (Hills and Sutherland, 1991), although the detailed spatial distribution of gainers and losers from these changes has received very little attention. This is a lamentable omission for, as we have argued in a previous paper (Martin *et al.*, 1992) there is almost inevitably a clear and geographical dimension to revenue-raising at the local scale. In this paper we will describe the use of a GIS for modelling the anatomy of the most recent UK local tax, the new council tax. Our GIS has been developed for an extensive area of Cardiff, Wales, which is known for planning purposes as the 'Inner Area' of the City. Our first objective is to predict the relative burden of the new tax upon all households within our case study urban area: we then conduct a preliminary sensitivity analysis in order to gauge the impact of possible city-wide over-valuation upon the incidence of revenue-raising. We will conclude with a discussion of the wider implications of this research for detailed GIS-based studies of the burden of local taxation.

The anatomy of the council tax

The introduction of the council tax in April 1993 will mark the end of a four- to five-year period during which the 'poll tax' or community charge has replaced the old domestic rates system (in 1990 in England and Wales, 1989 in Scotland). In 1991 the community charge was modified by increased levels of 'transitional relief' (targeted towards the households which lost the most through the initial change) and a blanket reduction in all personal charges: this latter change was funded by an increase in Value Added Tax (VAT) rates from 15 to 17.5 per cent. The shift towards increased VAT in 1991 was accompanied by the decision that central government was to bow to popular pressure and would replace the community charge with a 'council tax' with effect from April 1993. This tax is to be based upon the capital values of domestic dwellings as of April 1991.

It is tempting to caricature these changes as replacement of property-based taxation (the rates) with person-based taxation (the poll tax) and subsequent reversion to property taxes (the council tax). In practice, however, each has been a hybrid tax based upon mixtures of dwelling attributes and household characteristics (Hills and Sutherland, 1991; Martin *et al.*, 1992). For example, many households gained partial or even total exemption from the domestic rates on income grounds, whereas transitional relief by its very nature compensated for losses incurred as a result of historical rateable values. Nevertheless, a return in emphasis towards taxation based upon built form has undoubtedly occurred, and in this paper we will attempt to predict the consequences of this change in an inner city area. Household council tax bills will under certain circumstances be reduced: in particular, registered single person households will receive a discount, as will households that wholly (and in some cases partly) comprise registered full time students. Other discounts will apply to vacant dwellings and second homes. Local authorities will be required to allow for such discounts in setting the levels of their charges. In addition, households may be eligible for a range of means-tested rebates, the detail of which is not known at the time of writing. In this paper, we have been concerned only with the dwelling capital valuation component of the council tax, and we have not built in reductions for students and single-person households to our calculations of tax rates.

In devising an alternative vehicle for local taxation to the community charge, central government has been mindful of the pressures to devise a regime which has low implementation costs and which can be put into effect within a short time horizon. As a result, properties are to be allocated to one of eight bands, the widths and limits to which differ between England, Scotland and Wales. The fourth of these bands (Band D) is to attract a 'standard' council tax charge, with charges payable in the proportions 6:7:8:9:11:13:15:18 for the 8 bands A to H. The bands which are to be used in Wales are shown in Table 13.1, together with the ratios to the base (Band D) category. The spacing of these bandings is such that approximately 50 per cent of properties in each of England, Scotland and Wales should fall into the relevant base category (Anon, 1992). Capital valuations are intended to provide estimates of values as of 1 April 1991. Properties are being valued by a mixture of public and private sector valuers on the basis of external (front aspect) inspections only. In most cases, properties will not receive individual valuations, but rather will be based upon the value of one or more 'beacon' properties in each street. The capital valuation of such beacon properties will be premised upon the standard assumptions as to state of repair and availability for sale as set out in statute.

As with the domestic rates system, households will have the right of appeal against council tax valuations, and such appeals are to be lodged between April and November 1993. Substantial numbers of appeals are anticipated,

Table 13.1. Welsh property bands and relationship of each to base (band D) category.

Valuation Band	From	To	Relationship to Band D category
A	up to	£30,000	6/9
B	£30,000	£39,000	7/9
C	£39,000	£51,000	8/9
D	£51,000	£66,000	9/9
E	£66,000	£90,000	11/9
F	£90,000	£120,000	13/9
G	£120,000	£240,000	15/9
H	£240,000	and above	18/9

partly because of the process of inference from beacon properties to dwellings which have not been surveyed. Small area variation in the numbers and likely success rates of appeals are likely, not least because of wide variations in the levels of heterogeneity of built form at such scales. More generally, house prices have been falling steadily throughout most of Britain since April 1991, and thus households are likely to appeal against valuations based upon capital values which relate to a period close to the 1989 historic high of British house prices. Given the nature of the valuation process, and the general (but by no means universal) similarities between adjacent dwellings within the residential fabric, there is likely to be systematic patterning in the occurrence of properties that are misclassified into incorrect bands. In the following section, we develop a street-based GIS to model the likely bandings of properties within a large area of Cardiff, and then carry out a sensitivity analysis of the effects of mismatches between predicted and actual capital values in order to anticipate the effects of systematic overvaluation and/or successful appeals following the implementation of the council tax.

The Cardiff case study

Our empirical case study concerns the so-called 'Inner Area' of Cardiff, an area of predominantly Victorian and Edwardian housing which has been defined by the City Council for various urban renewal purposes. In more recent years, some local authority developments have occurred towards the edge of the Inner Area (Gabalfa and Tremorfa), whilst one area has been subject to redevelopment (Butetown) and others have experienced infill development. Like most areas of southern Britain, Cardiff has experienced a downturn in real house prices over the period 1989–92: however, for a variety of reasons house prices in this local housing market have remained fairly stable, with our own evidence suggesting fairly small absolute falls during 1991 and a continuing but small downward drift during 1992. The

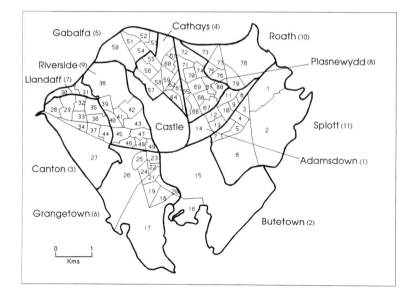

Figure 13.1. Community boundaries and HCS areas of Cardiff's Inner Area.

Inner Area breaks down into eleven whole or part Welsh Communities, and the Council in turn has broken this down into 81 'House Condition Survey' (HCS) areas (Keltecs, 1989). These HCS areas have been identified for housing policy purposes, and are defined as areas with considerable homogeneity of built form. The community boundaries and HCS areas are shown in Figure 13.1. Some 836 street segments can be identified within the area and these were digitized, in house, as part of a wider investigation of the redistributive consequences of successive local taxation regimes (Martin *et al.*, 1992). The Cardiff City Rates Register suggests that 47,014 hereditaments were located in these streets (some of which were split into separately rated units), and 45,658 of these were residential dwellings. We have entered into a spread-sheet information from the 1990 Cardiff City Rates Register, comprising street code, rateable value, and HCS area code for all the streets that lay within, or partly within, the Inner Area. Preliminary data cleaning was carried out on the spread-sheet. This was necessary because the council tax is likely to be slightly broader brush than the domestic rates in its implementation, in that hereditaments identified as rooms, bed-sits, garages and car ports in the rates register are very unlikely to attract separate capital valuations for the council tax. A series of text searches were therefore carried out in order to consolidate such properties into single addresses that can be identified in the rates register.

Next, a survey was carried out of the asking prices of houses for sale within the Inner Area, using advertisements in local newspapers and information obtained from estate agents' premises. A total of 796 asking prices were

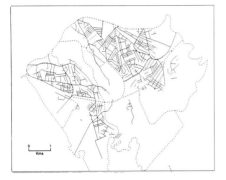 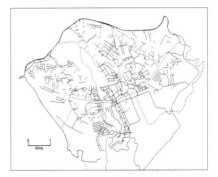

Figure 13.2(a). Spatial distribution of streets that feature at least one price survey dwelling.

Figure 13.2(b). Spatial distribution of streets that do not feature any price survey dwellings.

obtained, together with information as to property type, numbers of bedrooms, name and branch of estate agent through which property was offered for sale,[1] and whether or not the house price had been reduced since the property had first been put on the market. This represents 2.1 per cent of the total dwelling stock of the Inner Area. Asking prices were obtained for at least one property in 358 of the 836 streets: the spatial distribution of these streets is shown in Figure 13.2(a), while Figure 13.2(b) shows those streets for which no asking prices were identified. These maps present an inter-penetrating structure of streets with and without sampled dwellings, with the only concentrations of non-sampled streets corresponding to non-residential areas of the city: this evidence therefore suggests that a comprehensive coverage of dwellings within the Inner Area has been achieved. Exact addresses (i.e. street plus number) were identified for 232 properties in our house price survey, and the street name was the most detailed information that was readily available for the remainder. Further details of the survey, and its relationship to our broader research agenda, are given in Longley *et al.* (1993).

The 796 price survey dwellings were deemed 'beacon' properties for purposes of our analysis. Clearly they were available for sale on the open market, and their asking prices reflect private sector valuations of their worth. These properties were therefore used to estimate capital values for every hereditament in the Inner Area. An algorithm was devised in order to allocate values to all dwellings, in order that the results could then be manipulated and analysed within the GIS. This algorithm essentially assigned capital values using the following four-stage procedure:

1. The 232 addresses for which asking prices were known were assigned these values.
2. All addresses which were of the same construction type (e.g. 'house', 'flat', etc.) and lay in the same street as a price survey dwelling were assigned the asking price of that price survey dwelling. In a number of cases more

than one price survey dwelling was located in a given street: in such cases, the mean value of such price survey dwellings was assigned to the un-known rates register entries in that street.

3. Where no suitable property was located in the same street, a search was carried out for the nearest price survey dwelling of the same type in the same HCS area. HCS areas were deemed appropriate areal units to carry out imputation because of the generally homogeneous nature of their dwelling stock detailed above. As a diagnostic check to prevent compari-son with inappropriate dwellings, price survey dwelling values were dis-carded if the rateable value of the price survey dwelling fell outside of one standard deviation of the distribution of rateable values in the street for which capital values were to be imputed. Thus it was required that both dwelling type and rateable value were comparable before capital values were assigned. Where the precise rateable value of the price survey dwell-ing was not known (i.e. in up to 564 cases), the mean street rateable value was compared with the standard deviation of the mean street rateable values in the streets where capital values were to be assigned.

4. If none of these methods proved to be appropriate for the assignment of capital values, a regression model was used. Capital values were regressed against rateable values for the 228 price survey dwellings for which paired capital values and rateable values were known. For the Inner Area as a whole, the relationship was:

$$CVAL = 4003 + 358.64(RVAL) \qquad (1)$$
$$[1.29] \quad [18.03]$$

R-squared = 59.0%   R-bar-squared = 58.8%
[t-statistics in brackets] No. obs. 228

where CVAL denotes the capital value of a uniquely identifiable hereditament; and RVAL denotes the paired rateable value.

Exploratory analysis revealed that rather different parameter estimates were appropriate to the Roath community area: for details see Longley *et al.* (1992). Results from these regression analyses were used to estimate capital values for the remaining hereditaments.

Overall, 269 capital values were assigned using exact address-matching; 20,982 were assigned using a 'beacon' property from the same street; 22,523 were assigned using a 'beacon' property from a different street within the same HCS area; and 1884 were assigned using the results of the regression model. These methods are invariably imprecise and invoke assumptions of varying strength. However, a number of the principles that we have used (e.g. the assignment of 'beacon' properties and the use of rateable values to aid capital valuation in the case of 'non-standard' properties) have close counter-parts in the council tax valuation process. As such, we would contend that our GIS presents an alternative, and equally viable, means of ascertaining capital values for the Inner Area of Cardiff.

In a previous paper (Longley *et al.*, 1993) we have presented a map of banded valuations that result from implementation of this algorithm. For purposes of comparison with the results that follow in section 3 these maps are reproduced as Figure 13.3, which shows the modal council tax band for every street in the Inner Area. Figure 13.3 exhibits contiguity effects in the pattern of valuations, yet the size of our house price survey makes this unlikely that this could be simply an artefact of the data modelling method. Indeed, whilst at the street level the modal band accounts for over 90 per cent of valuations in 455 of the 836 streets, there are a considerable number of streets where this is not the case. This is shown in Figure 13.4. We have also reported that the predicted capital values are generally clustered quite closely about the modal band, with less than 5 per cent of all streets with over 50 per cent of dwellings that lie more than one band away from the modal band.

In absolute terms, the distribution of properties revealed in this exercise shows a concentration of property values in bands B, C and D, which is in broad correspondence with central government guidelines (Anon., 1992; see Table 13.2). We have used the figures in Table 13.2 to produce an estimate of the standard (Band D) charge which the City Council would need to levy in order to raise the same amount of revenue as that likely to be raised under the Community Charge in 1992/3.[2] Our calculations suggest a standard Band D charge of £234.82.

However, the allocation of properties to bands is likely to be very sensitive to changes in the valuations of the 'beacon' properties. At the time of the survey, most estate agents in our surveyed offices would tacitly admit that purchasers who were ready, willing and able to proceed with purchases would in most cases have been able to negotiate discounts on asking prices in the region of 5–8 per cent, and the reality may well have been that this was an underestimate. Various other sources of measurement error might also be mooted, such as the state of repair of owner occupied houses offered for sale relative to other properties (not least those in the private rental sector, which in the Inner Area of Cardiff is still dogged by problems of unfitness: Keltecs, 1989). From a policy standpoint, there might be sense in underestimating the capital values of properties that lay on the boundaries between adjacent capital value bands, in order to reduce the burden of appeals. Such a strategy might also have the effect of accommodating the effect of falling house prices although, at the time of writing, it is a moot point as to whether any such ploy would be able to mask the high absolute price falls that will characterise the period between valuation and instigation of the tax. For these and numerous other reasons, it is sensible to view these results as but one of a class of estimates, and to undertake a sensitivity analysis of them in order to anticipate a range of possible mitigating factors. We will now turn to consider this aspect of the analysis.

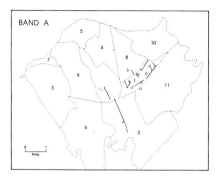
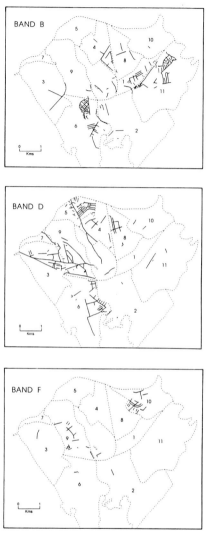
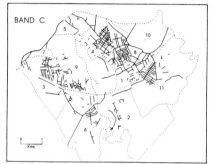
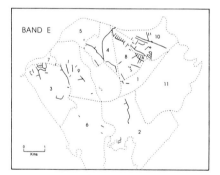
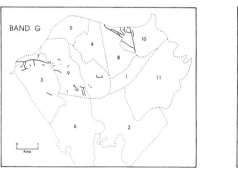
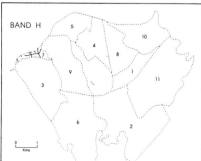

Figure 13.3. Assignment of streets to modal bands, based on price survey.

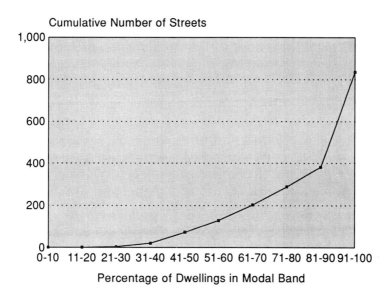

Figure 13.4. Distribution of dwellings about street modal bands in the Inner Area.

Table 13.2. Allocation of residential properties between council tax bands.

Valuation Band	Market valuations	% of r. props
A	2,092	4.6
B	8,226	18.0
C	16,047	35.1
D	8,053	17.6
E	5,252	11.5
F	1,641	3.6
G	2,030	4.4
H	2,317	5.1
non-residential	1,356	–
Total	47,014	100

Source: Authors, calculations

The sensitivity analysis

We believe that the database created for the Inner Area of Cardiff comprises the most comprehensive estimate of property values presently available for any substantial part of a British city (*The Independent*, 1992). Our survey was carried out in December 1991, and our belief is that the capital values obtained in our survey provide close approximations to the asking prices that were prevailing in April 1991, i.e. the base date to which the council tax

Table 13.3. Effect of a 10 per cent reduction in council tax valuations.

Valuation Band	10% discount valuations	% of resid. hereditaments	Change (%)
A	3,869	8.5	+3.9
B	12,772	28.0	+10.0
C	13,789	30.2	−4.9
D	5,907	12.9	−4.7
E	4,383	9.6	−1.9
F	756	1.7	−1.9
G	1,865	4.1	−0.4
H	2,317	5.1	0.0
non-residential	1,356	–	–
Total	47,014	100	0

valuations are to pertain. Moreover, our capital value modelling procedure is likely to be at least as sophisticated as that employed by central government, and our preliminary analysis of the results (Longley *et al.*, 1993) suggests that the distribution of capital values is intuitively plausible at both the inter- and intra-street scales. Our GIS provides a very flexible medium for the storage of this information, and it is therefore fairly straightforward to investigate a range of scenarios concerning validity and accuracy of our capital value information. In this section we will pursue one such scenario, namely the effect of reducing all of our valuations by a uniform 10 per cent: this scenario may be construed as correcting for systematic over-valuation by estate agents of realisable market prices, or as anticipating conservative valuation practices in the official valuation of hereditaments in order to minimise the likely magnitude of appeals, or some combination of these two factors.

The effect of this general reduction is to change the distribution of hereditaments laying within each of the bands, as shown in Table 13.3. Clearly any general reduction in values will have the effect of shifting properties that were near to the lower limit of a band into the next lowest band, and Table 13.3 shows that the effect of this is to cause significant increases in the number of properties (column 2) which fall into Bands B and A. This shift will cause a short-fall in revenue unless it is offset by an increase in the level of the standard (Band D) charge: invoking the same assumptions that were used in the previous section, we estimate that the standard charge for the Inner Area would increase from £234.82 to £245.55. This results in some significant savings by those households that move down one band, which are paid for by those that remain in the same bands: residents of hereditaments in the higher bands are proportionally more likely to remain in the same bands, with those in bands G and H least likely to shift between bands. These changes are summarised in graphical form in Figures 13.5 and 13.6. The mapped distribution of properties falling into each of the bands is shown in Figure 13.7.

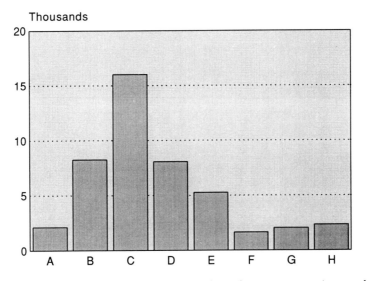

Figure 13.5(a). Allocation of properties to bands in the inner area using unadjusted price survey.

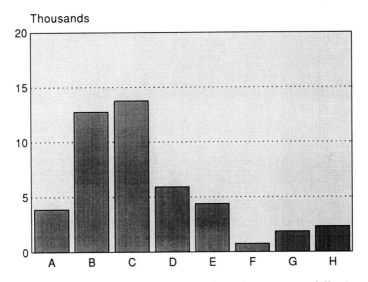

Figure 13.5(b). Allocation of properties to bands in the inner area following a 10 per cent reduction in prices.

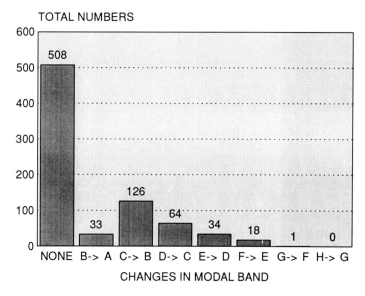

Figure 13.6. Changes in modal band distribution following 10 per cent price reduction.

Figure 13.8 shows the spatial distribution of streets where changes in property valuations cause the modal band to change. These maps exhibit concentrations of streets which cannot be explained by the spatial distribution of sampled dwellings alone (Figure 13.2). As such, they provide some evidence that appeals against valuations might be concentrated in small areas if our general scenario of systematic over-valuation were to be appropriate for the Inner Area as a whole.

Conclusions

We begin by restating our assertion that our exercise represents the first serious attempt to quantify the detailed incidence of the new council tax upon a local area. We feel also that it has demonstrated the use of a rudimentary spatial model in imputing capital values across a wide area and look forward to comparing our results with some of the official valuations when these are known. The use of a spatial model couched within a GIS environment presents an attractive alternative strategy for estimation of capital values. It is a particularly cost-effective exercise since our information has not necessitated primary data collection, yet it assimilates a wider range of dwelling attributes than do the actual official valuations themselves. Given that the official figures are not yet known, our study provides a rare application of GIS in a genuinely predictive context.

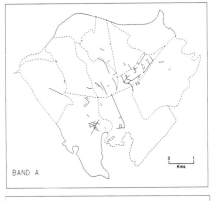

BAND A

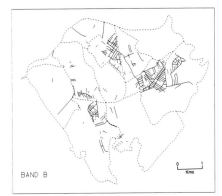

BAND B

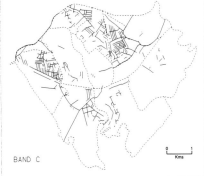

BAND C

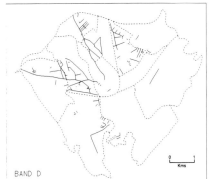

BAND D

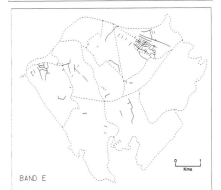

BAND E

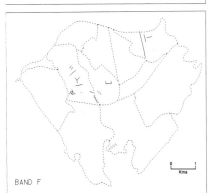

BAND F

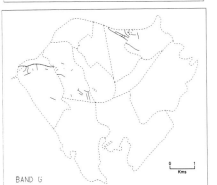

BAND G

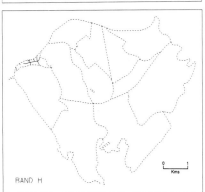

BAND H

Figure 13.7. Assignment of streets to modal bands, following a 10 per cent price reduction.

Figure 13.8. Spatial distribution of streets with changed modal price band.

The simulation might also be extended in order to incorporate further house price information (such as summary dwelling condition indices), and our model might be made more sophisticated in order to incorporate the length of time properties have been on the market, the apparent valuation policy of particular estate agents and estate agent branches, and whether or not a current asking price represents a reduction on a previous valuation.

Of more immediate concern, the flexibility inherent in storing our information within a GIS offers the possibility of analysing a wide range of scenarios concerning changing housing market conditions. We believe that our results are sufficiently robust to bear comparison with the official figures for the tax when these become available, and provide a 'first filter' for identifying those contiguous areas of the Inner Area where such appeals are most likely to prove successful. We have also provided information about the effects of revaluation upon the distribution of properties between bands, which has implications for the level of the standard charge that is to be set by local authorities. These and other topics will provide the focus to our future research.

Acknowledgement

This research was funded by the U.K. Economic and Social Research Council grant no. R000 23 4707.

Notes

1. The normal practice in Cardiff is for vendors to offer houses for sale through 'sole agency' agreements with estate agents, and a negligible proportion of houses in our survey are presumed to be on offer through multi-agency agreements.
2. Of course, the estimation of an equivalent Council Tax figure requires a number of assumptions at this stage: specifically, we assume that there will be no redistribution between the Inner Area and the remainder of Cardiff following introduction of the Council Tax, and that the amount of transitional relief channelled into the Inner Area remains constant. The first of these assumptions is unlikely, since the initial shift from the rates to the community charge has been calculated by Martin *et al.* (1992) to have led to an increase in the relative contribution of the Inner Area equal to about 4 per cent of total City revenues: if any shift following the introduction of the council tax were of similar magnitude our estimates would nevertheless still provide a rough guide. The second of these assumptions is unverifiable until central government reveals its plans for transitional relief arrangements.

References

Anon, 1992, The Council Tax, *Estates Gazette*, 153–154, 7 March.

Hills, J. and Sutherland, H., 1991, Banding, tilting, gearing, gaining and losing: an anatomy of the proposed Council Tax, *Welfare State Programme Discussion Papers*, No. 63, London School of Economics, London.

Keltecs, 1989, *Cardiff House Condition Survey. Phase 1: Inner Area Final Report*, Keltecs (Consulting Architects and Engineers) Ltd., Grove House, Talbot Road, Talbot Green, CF7 8AD.

Longley, P., Martin, D. and Higgs, G., 1993, The geographical implications of changing local taxation regimes', *Transactions of the Institute of British Geographers*, **18**, 86–101.

Martin, D. J., Longley, P. and Higgs, G., 1992, The geographical incidence of local government revenues: an intra-urban case study, *Environment and Planning C*, **10**, 253–265.

The Independent, 1992, 'Rich "will gain most from the council tax", 17 August, p. 2.

Index

adjacency analysis 165
ANNEAL 223, 234–44
  TransCAD 236–4
  creating files 240–2
  results 243–4
annealing *see* simulated annealing
applications generator 181–3
arcgraph 153
ARC/INFO 6, 60, 105
  areal interpolation 130, 135, 143
  object oriented analysis 173, 177
  population density modelling 7,
    189–217
    Buffalo 204–17
    integrating software modules
      199–204
    problems 195–9
  spatial point process modelling
    150–1, 153–7
  statistical analysis 14–15, 24, 27,
    29–30, 32–5
ARC macrolanguage (AML) 24
  population density modelling 191,
    193–4, 204
  spatial point process modelling 151,
    155, 157
ArcPlot 193–4, 200, 208, 209
area data statistical analysis 17–20
areal interpolation 6, 16, 121–44,
  150
  ancillary data 124–30
  binomial 123, 126–8, 133–4, 144
  boundaries 121, 125, 132, 143
  continuous case 134–9
  intelligent 143–4
  operationalizing 130–1
  Poisson 123, 126–7, 131–4, 144
  terms and notations 122–3
  weighting 123–4
artificial life (AL) 94–7, 102
attribute data 107–9, 147
  analysis modules 45, 48–9, 58–60
  pattern analysers 83–103
    tri-space 83–4
    tool design 85–90, 102
  population density modelling 191,
    195, 199
autocorrelation 5, 18–19, 29–32, 37,
  71–2, 74
  cancer analysis 51, 57, 58, 60

spatial point process modelling
  149–51
urban analysis 193
autocorrelograms 29
autoregression 107

Baye's Theorem 28
Bayesian smoothing 18, 27, 28
Besag's sequential Monte Carlo test 94
binomial areal interpolation 123, 126–8,
  133–4, 144
bootstrap 4, 32
boundaries 57, 248
  areal interpolation 121, 125, 132, 143
  Cardiff council tax 264, 267
  edge correction 20, 26–9
Bristol 253, 258
Buffalo 7, 192, 199–200, 204–17
  running software 207–15
buffering 130, 165

C (language) 24, 150, 238
  population density modelling 191,
    193, 204
California 109–10
cancer 91
  mortality rates 46, 50–61
  raised incidence 155–6, 158
  Sellafield 102
canonical correlation 18, 34
car ownership 133–4, 247
Cardiff
  council tax 8, 261, 263–75
    sensitivity analysis 269–72
  social indicators 253–7
CART 24
cartographic algebra 15
CHEST deal 151
class inheritance (OOP) 168
clustering 6, 67–70, 74, 108, 160
  analysis 17–18, 26, 34
  cancer 51, 53
  house prices in Cardiff 267
  leukaemia 149
  pattern analysis 94, 101–2
  social indicators 251, 253
coastal vegetation in California 109–10
coefficient of determination 201
competing destination (CD) models
  72–3

contiguity
  council tax in Cardiff 267
  redistricting 223, 224, 226–31, 233,
    235, 241–5
core code (OOP) 168–71, 175–6, 181,
  183
correlation coefficient 57
correlograms 18, 30, 31
cost minimization 183–4
council tax 8, 261–75
  sensitivity analysis 269–72
coupling 22, 24, 148–50, 193
covariance structure 29–32, 36
crime pattern analysis 97–100

dasymetric mapping 124–5
database anomalies 83, 89
data transformation 22–5
descriptive analyses 22–5
deviance 133–4, 144
discretization 47, 48, 50
diseases 85, 87, 91, 149, 153, 193
  cluster analysis 17, 26
  spatial point process modelling 152,
    155–8
  see also cancer

edge correction 20, 26–9
EM algorithm 126, 127, 129–30, 131,
  139
enumeration districts (EDs)
  cancer 51–5
  leukaemia 91
  social indicators 248, 250
entropy maximization 222
epidemiology 85, 87, 91, 149, 153, 193
  cluster analysis 17, 26
  spatial point process modelling 152,
    155–8
  see also cancer
ERDAS 34
errors 16, 48, 85, 90, 151, 267
  cancer analysis 51, 56
  statistical analysis 31–2
  surface models 250, 252, 255
exploratory data analysis (EDA) 74

facility location 222
food poisoning 153
FORTRAN 6, 200, 208
  population density modelling 191,
    193, 204
  spatial point process modelling 153,
    159

statistical analysis 24, 27, 29
TransCAD 238, 240–2, 243
Fotheringham measure 72

GENAMAP 24, 35
general linear modelling (GLM) 33,
  37
Geographical Analysis Machines
  (GAMs) 87, 88, 91–2, 148–9,
  156
geometric pattern analysis 74
geostatistical econometric modelling
  31–3
GLIM 24, 30, 32, 33, 150
  areal interpolation 130–1, 143
  pattern spotting 89
  urban analysis 193
Graphical Correlates Exploration
  Machine (GCEM) 92
GRASS system 24, 192
gridding 6, 14, 17

health provision 247–8, 253, 258
heterogeneity 57, 105–19
  descriptors 107–9
  pixel data 109–15
heteroskedasticity 105, 107
histograms 150
housing
  prices 8, 134–42, 261–75
    anatomy 261–3
    Cardiff 263–75
    Preston 134–42
    sensitivity analysis 269–72
  social indicators 247–8, 253, 256

IDRISI 24, 26, 30, 32, 105, 149, 193
inference 83
  SDA modules 52–5, 58
INFO-MAP 25–6, 28, 30, 32, 34,
  149
intelligent areal interpolation 143–4
intelligent knowledge based systems
  (IKBS) 90
interaction data 17–18
inverse power function 195–8
Iterated Conditional Modes (ICM)
  28

K-functions
  spatial point process modelling 152,
    153–4, 158–9
  statistical analysis 18, 20–1, 25–7, 29,
    36–7

kernel density estimation 18, 20–1, 27, 29, 107
  spatial point process modelling 155–6, 158
  surface model 250–2
kernel regression 18, 28
kernel smoothing 20, 27–8, 158–9
Knox's space-time statistic 94
kriging 14, 18, 20, 29, 31–3, 36–7

languages 4, 166
  spatial point process modelling 149, 151, 157–8
  statistical analysis 14, 21, 24–7, 38–9
  TransCAD 238, 240
  see also macrolanguages
leukaemia 91, 149, 155, 158
linear regression 149, 197–9
location/allocation modelling 14, 15, 35, 193
locational data 17–20, 65, 147
  modules 45, 58–60
loglinear regression 197–9

macrolanguages 105, 130
  urban analysis 190–1, 194
  see also ARC macrolanguage
mapping packages 25
marked point process 17
Markov random field 28
maximum likelihood 198–9
McDonald's customer spotting 76
menus
  population density modelling 200–4, 211
  TransCAD 237, 239–42
message passing (OOP) 167–8
Metropolis algorithm 233–4, 236
mikhail 153, 155
MINITAB 24, 31, 32, 56
missing value interpolation 16, 40
modifiable areal units 19–20, 74
Monte Carlo tests 6, 92, 93–7, 99–100
  procedure 94
multiobjective programming 65
multivariate data 17–19, 26, 29–31
  SDA modules 47–50, 53, 56–8, 61
  techniques 33–4, 37
  tri-space analysis 86, 88–9

nearest neighbour 75
  spatial point process modelling 152, 158
  statistical analysis 18, 25–7, 30, 32

negative exponential function 195–8
network analysis 15
neurocomputing 90
non-stationarity 9, 105

object class 48, 166–7
object oriented programming (OOP) 7, 165–84
  concepts and terms 166–70
  generic LP classes 170–1
  South Africa 176–81
  TRAILMAN 183–4
  transportation 171–3, 174–6
operationalizing 130
optimization modelling 7, 190, 221–45
  ANNEAL redistricting algorithm 234–6
  redistricting problem 223–32
  simulated annealing 232–4
  TransCAD-ANNEAL 236–44
outliers 9, 32, 215
  cancer study 55–6, 59
overcrowding 133–4, 247, 253, 257
overlays 130, 165, 194
  polygon 16, 121
  spatial point process modelling 148, 149
overloading (OOP) 168

PASS 130–1, 150
pattern
  analysers 83–103
    artificial life 94–7
    GIS databases 84–5
    results 97–101
    STAM 91–4
    tri-space analysis tools 85–90
  analysis 5–6, 13, 15, 17–18, 34, 67, 74–5
    clustering 94, 101–2
    crime 97–100
    description 45
    proximal data 115
    recognition 5–6, 54, 66–7, 148–151
    spotting 83–5, 88–90, 95
    STAM 94–100
pixel data 109–18
Planning Support Systems 190
point data process 17
point pattern analysis 5–6
Poisson 57, 93, 152
  areal interpolation 123, 126–7, 131–4, 144
polygon overlay 16, 121

population 8
  areal interpolation 124–8, 131–4
  binomial case 134
  density modelling 7, 189–217
    Buffalo 204–17
    integrating software modules
      199–204
    problems 195–9
  epidemiology 152
  equality 223–4, 230–1, 233, 243
  mobility 57
  Poisson case 131–3
  social indicators 247–58
    surface models 249–52
postcodes
  cancer study 51–3
  epidemiology 152
  house prices in Preston 135–42
  surface models 249–53, 256
proximal space databases 105–7, 115

quadrat analysis 75, 149, 152
quadtree storage system 109
quenching 232, 235

raisa 155–6
raster data 17, 109, 149, 192
  surface models 251–8
reality, views of 48–50
redistricting, political 8, 221–45
  ANNEAL algorithms 234–6
  improvements 244–5
  mathematics 223–32
  simulated annealing 232–4
  solution tree 225
  TransCAD-ANNEAL 236–44
REGARD *see* SPIDER
regression 266
  kernel 18, 28
  linear 149, 197–9
  loglinear 197–9
  modelling 56–8, 61
  spatial 31–3, 36, 37
routing 15

S (language) 21
SALADIN 35
SAS 24, 31, 32, 34
scale effects 6, 86, 115
scatterplots 23–5, 150
  SDA modules 53, 55–6, 58
search parameters, 93, 98–9
seeding procedures 224
semi-variogram parameters 107

sensitivity analysis 269–72
simulated annealing 8, 222–3, 228,
  232–6, 242
site selection 76
smoothing 18, 20, 27–9, 36, 37
  Bayesian 18, 27, 28
  kernel 20, 27–8, 158–9
  proximal databases 107
  spatial point process modelling
    155–6, 158–9
social indicators 247–58
  Cardiff 253–7
  surface models 249–52
socio-economic variables
  areal interpolation 121, 123
  cancer study 51, 53, 55, 56
South Africa 176–81
SpaceStat 31, 32
space-time-attribute analysis machine
    (STAM) 91–4
  pattern search 94–100
  see also STAM/1
space-time-attribute creature (STAC)
    96–101
SPANS 24, 34, 35, 77, 105
Spatial Analysis Module (SAM) 7,
    193
Spatial Data Analysis (SDA) modules
    5, 45–62
  conduct 52–5
  design 50–61
  requirements 55–8
spatial database establishment 13
spatial dependence 105–19
spatial econometric modelling 31–3
spatial interaction 5–6
  modelling 195, 238
  models 34–5, 37
  theory 70–1, 74, 88
spatial point process modelling 6,
    147–61
  existing approaches 149–51
  proprietary GIS 151–7
  statistical programming 157–60
spatial regression 31–3, 36, 37
spatial video modelling 27
SPIDER (now REGARD) 25, 27, 28,
    31, 39
SPLANCS 6, 25, 27, 158–9
S-Plus 6
  spatial point process modelling 151,
    154, 157–60
  statistical analysis 21, 24–5, 27, 29,
    38–9

SPSS 24, 31, 56
start-up costs 45
stationarity 108
Statistical Analysis Module (SAM) 24, 30, 150
statistical spatial analysis 13–40
  potential benefits and progress 21–36
  useful techniques 15–21
STAM/1 92–3, 97
  crime patterns 98–9
summarization 15–16, 21, 23–5
SUN platform 199
SUN SPARC 151
Sunview 200, 204, 207, 208
surface modelling 8, 247–58
  social indicators 252–7

Thiessen polygons 148
three dimensional modelling 15
TIGER 189, 191, 205–6
time-series analysis methods 88
TRAILMAN 183–4
TransCAD 8, 35
  ANNAEL 236–45
  FORTRAN 238, 240–2, 243
  population density modelling 190, 192, 193
  redistricting 222, 236–45
transportation 15, 67, 221, 238
  OOP 171–3, 174–6
  South Africa 177–81
travelling salesman 222, 238
trend 108–9
trend surface analysis 18, 149
TRIPS 35
tri-space analysis 83–4
  tool design 85–90, 102

unemployment 253, 255
UNIGRAPH 153
UNIMAP 31
UNIRAS 24, 32
univariate data 18, 27, 31
  SDA modules 47–50, 53, 55–8
urban land use 221
urban planning 106

variance 107–8, 110–19
variograms 18, 20, 25, 28–31, 37, 150
views of reality 48–50
virtual functions 168
visualization
  geographical 14, 36
  statistical results 22–3, 26–7, 34

weighting
  areal interpolation 123–9, 133, 138–40, 144
  surface models 251–2
weights matrix 60–1
windows
  population density 199–204, 208–15
  spatial point process modelling 150, 158–9
  statistical analysis 23, 24–6, 30, 38–9
  TransCAD 237–8, 240–2

zones 8–9
  areal interpolation
    source 6, 122–31, 133, 143
    target 6, 122–31, 133, 139, 142–4
  social indicators 248–9, 252–3
zooming 21
  population density modelling 202–4
    Buffalo 207, 210–11
  TransCAD 237